USING BUSINESS STATISTICS
A Guide for Beginners

Terry Dickey

A FIFTY-MINUTE™ SERIES BOOK

CRISP PUBLICATIONS, INC.
Menlo Park, California

USING BUSINESS STATISTICS
A Guide for Beginners

Terry Dickey

CREDITS
Editor: **Christopher Carrigan**
Managing Editor: **Kathleen Barcos**
Typesetting: **ExecuStaff**
Cover Design: **Carol Harris**
Artwork: **Ralph Mapson**

Copyright 1994 Crisp Publications, Inc.

Printed in the United States of America

English language Crisp books are distributed worldwide. Our major international distributors include:

CANADA: Reid Publishing Ltd., Box 69559-109 Thomas St., Oakville, Ontario Canada L6J 7R4. TEL: (416) 842-4428, FAX: (416) 842-9327

AUSTRALIA: Career Builders, P. O. Box 1051, Springwood, Brisbane, Queensland, Australia, 4127. TEL: 841-1061, FAX: 841-1580

NEW ZEALAND: Career Builders, P. O. Box 571, Manurewa, Auckland, New Zealand. TEL: 266-5276, FAX: 266-4152

JAPAN: Phoenix Associates Co., Mizuho Bldg. 2-12-2, Kami Osaki, Shinagawa-Ku, Tokyo 141, Japan. TEL: 3-443-7231, FAX: 3-443-7640

Selected Crisp titles are also available in other languages. Contact International Rights Manager Suzanne Kelly at (415) 323-6100 for more information.

Library of Congress Catalog Card Number 93-72980
Dickey, Terry
Using Business Statistics
ISBN 1-56052-250-6

This book is printed on recyclable paper with soy ink.

ABOUT THIS BOOK

Using Business Statistics can be used effectively in many ways. Here are some possibilities:

Individual Study. All you need is a quiet place and some time. You can understand this book with only your own resources, and it provides valuable feedback and practical ideas to improve your knowledge and skill.

Workshops and Seminars. Leaders can ask participants to read this before a workshop or seminar. When they know the basics, participants should find the experience more valuable; more time can be spent on concept extensions and applications. The book is also effective when trainers distribute it at the beginning of a session and lead participants through the contents.

Remote Location Training. The book is useful for employees who cannot attend home-office training sessions.

Informal Study Groups. It is also ideal for informal group study. The format is brief, easy to follow, and inexpensive.

Statistics Classes. Because of its intuitive approach, this book makes an ideal supplement to more rigorous statistics texts. Undergraduates will find it useful as they complete their first research projects.

Business Applications. The study of statistics is fundamental to topics such as quality assurance, process control, market research, and human resources management. This book is an excellent way to upgrade statistical skills of participants in business training programs without diverting them from their main focus.

There are other possibilities that depend on your objectives, programs or ideas. Even when you finish, this book will still be useful as an easy way to review.

ABOUT THE AUTHOR

Terry Dickey, author of *The Basics of Budgeting* and *Budgeting for Your Small Business*, has helped business executives apply statistical analysis to the real world for a number of years. He holds a master's degree in business administration and has held responsible planning positions with Pizza Hut Inc., Rent-A-Center Inc., and Safelite Glass Corporation. He has also taught business research to adult, nontraditional college students and coached them through statistical applications in their first research project. He understands the special needs of students without prior experience in this sometimes complex field. He is currently a management consultant based in Wichita, Kansas.

To the Reader

Are you intimidated by statistics? If so, you are not alone. Many college students and business executives feel the same way.

Experience in counseling both groups has convinced me that, although sometimes intimidating, statistics is a very valuable skill. Students and managers who can use statistics see issues more clearly. They ask better, more important questions. They solve problems faster and more easily. In the modern world of management by the numbers, exploding information and global competition, there is simply no substitute.

This book is an introduction to this important skill. Master it and you will never find statistics intimidating again.

Terry Dickey

CONTENTS

Introduction

Tracy just received surprising news from his supervisor. "Tracy, the company has been sold and I am retiring. These new people are strictly 'by-the-numbers.' They are reasonable enough, but they study everything. Don't even talk if you can't tell them 'how much.' Things are going to change."

"They certainly are," Tracy thought. "What in the world am I going to do?"

What would you do?

AN IMPORTANT TOOL OF SUCCESSFUL PROFESSIONALS

Tracy has discovered a brutal fact: In most organizations, if you can't speak statistically, no one listens. Like many students, managers and other professionals, Tracy knows that a working knowledge of statistics will improve his odds of career success.

College students study statistics in management, marketing, health-related professions, economics, psychology, sociology, education and many other programs. It is impossible to achieve professional status in almost any field without a working knowledge of statistics.

Business owners, managers and executives use statistics to analyze companies, to uncover emerging trends and to make decisions. Even nonmanagers use statistics when measuring quality, performance, etc. Statistics is an important tool for *everyone* in the modern company.

Professionals in almost all fields must understand and discuss complex ideas. They must deal confidently with risk. They must act on faint signals buried beneath a daily avalanche of data. Because it is so appropriate for these tasks, statistics has become a vital part of professional competence.

Warning: You should not undertake any form of business, personal or financial risk on the basis of this material. Although this book will increase your skill, it will not make you an expert. So, if you need to use statistics to make a decision that involves risk, *seek competent professional counsel to review your plans and to confirm your results*. Often statistical advice is easily and inexpensively available from consultants, from more experienced managers or from the faculty at a local university.

SOME USEFUL TIPS

This book is an introduction to statistics. Upon completing it, you should be able to understand what statistics is all about; apply normal curves, regression and cross-tabulation to actual problems; avoid common sources of error; and understand statistical symbols and jargon. Although not an exhaustive text, it will provide a good base for additional study and advanced applications.

Here are some ideas that will help you use this book effectively. Before starting, you should understand very basic high-school algebra. If you are fuzzy about square roots and simple equations, brush up a bit before you start. Also, read the sections in order, since each builds on earlier ones, and be sure you understand each main idea before proceeding to the next. Don't skip the exercises; although simple, they will make important ideas clearer. If part of the book seems hard to understand, try reviewing earlier sections for ideas you may have missed.

You can use the information in this book whether you do statistical analysis by hand, or use statistical or spreadsheet software. If you have access to statistical software, this book will show you how to create a data file for the program. You will also find it easier to select which procedures to run, and you will gain insight into how to interpret some of the results.

You can also do a great deal of statistical analysis with spreadsheet software. In fact, all of the problems in this book can be used to create simple, general-purpose templates. Once they are created, they will let you analyze many different kinds of statistical problems in seconds. Such spreadsheets are a handy place to keep notes about different statistical techniques. Also, spreadsheets can grow with you as you master additional techniques from other sources besides this book.

. . . Tracy left the conversation deep in thought. He had not reached his present level in the company by waiting for things to happen. "Time to brush up my skill in statistics," he thought. He was right . . .

I

How to Turn Questions into Measurements

SECTION 1

Determine If Research Is Appropriate

Angelo, the plant manager, was meeting with his shipping manager, Clark, about problems with late shipments. Sales were great, but the orders came in waves that first swamped the shipping department, then left it with nothing to do.

"You need to have the inventory clerks fill orders when shipping is really active," Angelo offered. "It just seems like the right thing to do."

Clark was unconvinced. "I want to use statistics to identify groups of parts that we could assemble on slow days. Then, on busy days it will be much more efficient to assemble these groups of parts into a complete order. I want the inventory clerks to do the statistical work, it will take six months to develop, and we will probably need to hire a couple of consultants. But when it works, it will solve the problem permanently."

Angelo frowned. Could he afford six more months of problems to find out?

What would you do?

DECISION-MAKING STRATEGIES

How do you make decisions? Most people use a mix of three approaches:

Judgment

When a speeding car bears down as you cross a street, a snap judgment can save your life. Many situations that require *application* of knowledge (day-to-day business decisions, clinical medicine, etc.) use judgment as the primary management skill.

Paralysis

Sometimes, for personal or political reasons, a conclusion is the last thing one wants to reach. You may find an occasional, devious use for "studying an issue to death."

Research

Decisions with high risk of unfortunate consequences (and with time and resources to reduce that risk) call for research. Testing a new drug is very different from emergency room medicine. Changing strategic direction, evaluating securities, or launching a new product are similar situations in business.

The only difference between these three approaches is the amount of time allowed for research.

Judgment **allows little or no time.**

Paralysis **requires an unlimited amount.**

Research **(and the use of statistics) lies somewhere in between.**

THE SIX HURDLES TO USING RESEARCH

Since going on vacation achieves the same result as paralysis (but with greater pleasure), the real choice is between judgment and research. The preferred choice is always judgment, unless you can answer "yes" to the following six questions:

1. IS THE ISSUE IMPORTANT ENOUGH? If the cost of a bad decision is small, research is not worth the effort.

2. CAN THE RESULTS INFLUENCE A DECISION? What would you do based on judgment alone? Is there a realistic chance for research to change your mind?

3. IS EXISTING INFORMATION INADEQUATE? It is usually faster and cheaper to ask someone who already knows. Has someone else already developed a good solution?

4. IS THERE ENOUGH TIME? Great studies are worthless if results arrive too late. Sometimes delay is worse than a bad decision.

5. IS THERE ENOUGH MONEY? Although a study might save a great deal of money, it is worthless if an organization can't fund it.

6. IS THERE ENOUGH SKILL? One company spent a year collecting a copy of every piece of paper produced in field offices. The company pulled in managers to code this data, hired temporaries to tabulate it, and created a custom database to analyze it. Everyone involved thought it was a great idea—until the very end, when no usable information was produced. Are the skills available to properly plan, control and execute *your* project?

Failing *any* of these hurdles means you should forget research. Instead, use your best judgment and whatever information is already available.

Exercise #1

1. List the six hurdles all research projects must clear before they can be considered justified.

    a. _____

    b. _____

    c. _____

    d. _____

    e. _____

    f. _____

2. Describe a potential research problem. Examine it using the six hurdles.

    Problem: _____

    a. _____

    b. _____

    c. _____

    d. _____

    e. _____

    f. _____

Summary

Before starting a research project, be sure you have adequate time, money and expertise to produce good results. The question should be important, and existing information should be inadequate. Also, there should be a realistic chance that the results of the research will actually influence the decision.

> . . . Angelo looked squarely at Clark. "We need to solve this in four weeks, not six months. Also, I doubt whether the inventory clerks have the expertise to develop the right statistical models. As for consultants, you know how tight the budget is; there's just no money and I doubt that the ideal solution will save that much more. Let's make a judgment call and use the clerks to pull orders for right now. Perhaps later, we can come up with a faster, cheaper way to do the project.". . .

SECTION 2

Create a Model
of the Research Problem

Toni was a corporate troubleshooter. Her latest problem involved a manufacturing plant where training expenses were too high and productivity was declining. She traced the problem to excessive employee turnover. Conversations with employees and managers were inconclusive; yet, something had to be done.

"Time for some research," Toni thought. "Where should I start?"

Where would you start?

MODELS

To use statistics effectively, you must understand *modeling*. To understand almost *any* complex thing, all of us visualize, create and study models: mental images that are *similar* to, but *simpler* than the real world. Building such a model is Toni's next task.

Models are very important to all professionals. Engineers "fly" model aircraft in wind tunnels. Pilots use flight simulators. Executives try out decisions on spreadsheets. A college education itself is really nothing more than learning basic models: supply and demand (economics), the time value of money (finance), operant conditioning (psychology), relativity (physics), etc.

The goal of all models is to produce simple rules for a complex world. As our understanding grows, those rules change along with the models that produce them. One hundred fifty years from now, ordinary medical treatment will be better than today's best, due to a better model of health care.

The main function of statistics is to build, apply, test or modify a model. The next few pages explore the relationship between statistics and modeling, and describe the first three steps of statistical analysis.

Exercise #2

1. Name three models that are important at work or school.

   _____

   _____

   _____

2. Are you currently using versions of those models that are different from the ones you first used? How are they different?

   _____

   _____

   _____

DATA, CONCEPTS AND THEORIES

Models are created from data, concepts and theories.

DATA

Data are measurements, records of events that are hopefully accurate and bias-free. Ideally, someone else at another time and place could create similar data by using the same procedures. Examples of data: a list of IQ scores, the responses of customers to a survey, output figures for a production process.

CONCEPTS

Concepts are abstract ideas that are relevant to the research problem. Examples of concepts: intelligence, customer satisfaction, quality. (Notice how they correspond to the data examples in the previous paragraph.) Since abstract ideas cannot be measured, concepts are always *operationally defined* as some type of data. An *operational definition* (of a concept) is data that is assumed to adequately measure the concept for the purposes of a research study. The same data can operationally define many concepts. For example, *number of defects* could be the definition of *quality*, *supervisor effectiveness*, *maintenance*, etc., depending on the needs of the study. Also, a single concept can be operationally defined many ways. *Quality* might be operationally defined by *number of defects*, *customer reports*, *tests of individual parts*, or by many other measurements. The important issue is whether the operational definitions adequately represent the concepts.

THEORIES

Theories explain the *relationships* between concepts. "Employee motivation increases with compensation and intangible rewards" is an example of a theory that suggests a relationship between three concepts. (Can you name them?) A simple theory can also be called a *hypothesis*. Almost all projects that require statistics start with a theory. For example, Toni might propose that employees have been leaving her firm because of problems with *hours*, *pay*, *supervision* and *working conditions*. By stating a theory, Toni has structured her analysis:

- She has identified key concepts to examine more closely.

- She has broken a difficult problem into smaller, more easily understood pieces.

- She has a way to set priorities and direct future activity.

FUNCTIONS AND VARIABLES

Functions are an efficient way to state theories. Here is Toni's. It reads, "Turnover is a function of hours, pay, supervision and working conditions."

turnover = f (hours, pay, supervision, working conditions)

To operationally define *working conditions*, Toni might use *average decibel level on the shop floor*, or *hours between report time and 8:00 a.m.*, or perhaps *the number of hazardous chemicals used*. She might even use all three.

Exercise #3

1. Can you suggest other operational definitions of working conditions?

_____

_____

2. What would you suggest as good operational definitions for the other concepts (hours, pay and supervision)?

_____

_____

Variables are elements of functions that can change or "vary." *Independent variables* determine the values of other variables. In this example, *hours, pay, supervision* and *working conditions* are independent variables. *Dependent variables* are those whose values flow from, are associated with or can be predicted by values of the independent variables. They *depend* on them. In this example, *turnover* is a dependent variable. The word *variable* is often used loosely. When you hear it, think carefully about whether the speaker means the concept itself or an operational definition of it.

Exercise #4

Think about a problem you have recently faced. Write down several different theories to model it. Write them as functions using the *turnover* function as an example. Circle the independent variables. Suggest operational definitions for each variable.

_____

_____

_____

THE FIRST THREE STEPS

The information you have just read may seem a bit theoretical, but it has a very practical pay-off. Three steps are always necessary to prepare a problem for statistical analysis.

STEP 1. Define the problem as a theory using specific concepts (or specify a single concept to measure). Write it down as a function with a variable for each concept.

STEP 2. Review each variable to be sure the function properly reflects the dependent/independent relationship.

STEP 3. Define each variable operationally.

Exercise #5

1. Define a possible research problem as a theory, using specific concepts. Write it down as a function with a variable for each concept.

_____

_____

_____

2. Classify each variable as independent or dependent. (Circle the independent variables.)

3. Operationally define each variable. Are there other operational definitions that might be better measurements of the concept?

_____

_____

_____

_____

4. All of the sections in this book start with an introductory story. If time permits, read each one now. Apply the three steps to each story, just as you did above. Write your answers in the margin of the first page of each section.

Summary

The use of statistics almost always involves creating, modifying or challenging a model. A model is a mental image of reality that is similar to but simpler than the real world. Most models can be described as functions. Variables are elements of functions that assume different values and are either independent or dependent. You should define your research problem as a theory that relates important concepts to one another, and write it down as a function with variables for each concept. Classify the variables as dependent or independent, and create operational definitions for each. By carefully thinking about the problem this way, you build a strong foundation for statistical analysis.

. . . Toni thought, "I think employee turnover is a function of hours, pay, supervisory skill and working conditions. I can get information on hours and pay from the payroll department. I can measure supervisory skill with ratings from the plant manager, and I trust my judgment to measure working conditions. Now at least I know where to start." . . .

SECTION 3

Organize the Data

Jim sold physical education equipment to school districts. His sales manager had placed him on special assignment to increase sales of a new line of equipment. The equipment really improved physical fitness, but funds were tight and schools were hard to convince. Several months ago, another sales representative had loaned some of the new equipment to an important school to evaluate. If Jim could prove the equipment increased physical fitness, the district would finalize that sale and probably buy more. Also, a successful sale to that district would make sales in other districts much easier. The district had agreed to supply whatever information Jim needed, and he had convinced an analyst at headquarters to analyze whatever the district provided. "Just bring me a computer file with the data," the analyst had said. "I'll meet with you at headquarters early next week."

Now the school was on the phone asking what information he needed. To Jim, the tough part was figuring out the next step.

What would you do?

THE BASIC TYPES OF DATA

Like all projects that involve statistics, Jim's analysis requires collecting and organizing raw data. After he creates operational definitions, Jim will know what information to collect from the school. However, to be ready for the analyst, he must also organize it into a data file.

A data file includes not only numbers, but also a definition of the *type* of data each number represents. It is important to understand the type of data because it partly determines the statistical tools you may use. There are four of these basic types of data: nominal, ordinal, interval and ratio.

Nominal

Nominal data uses numbers as names, for classification purposes only. For example, Jim might identify three physical education classes as numbers 1, 2 and 3. He might also record another nominal variable called *use_equip*, to indicate whether a gym class had access to the equipment (1=yes, 2=no).

Nominal variables are useful for identifying groups within the data, such as gender, department number, experimental vs. control group, and so on. Grouping data helps the analysis of other variables, as well. For example, Jim might want to measure physical fitness somehow, and then compare the average for those who used the equipment to those who did not. To do so, he would need to record *use_equip* (a nominal variable) and a measure of physical fitness for each student.

He could also use another nominal variable to measure physical fitness, perhaps by recording the coaches opinions for each child (1=yes they are fit, 2=no they are not).

However, in spite of their usefulness, nominal data contain too little information to use with many powerful statistical techniques—even something as simple as the average. For example, if Jim were to record class numbers for every child and then average them, the result would be meaningless.

Ordinal

Ordinal numbers are rankings—first, second, third and so on. For example, to measure physical fitness, Jim might select the finish order in a race. Such a ranking would be ordinal data. It would tell Jim more than nominal data (fit or unfit), but less than if he actually measured the time in the race.

Ordinal data are useful in situations where it is difficult to get better measurements. For example, the gym teachers might have trouble saying "Zeph is *twice* as fit as Asher," but they might be comfortable saying "Zeph is *more* fit than Asher" (and less than Cory, etc.). Ordinal data still contains too little information to use statistics based on the average, but it does allow use of more powerful tools than does nominal data. For example, you could pick out an individual in the exact middle of the rankings and have some basis to represent him or her as the typical student.

Interval and Ratio

These two types of data are similar. Interval data means that you can measure the size of the difference between numbers. Not only do you know that Lauren is "fast" and Jo is "slow" (nominal measurements) or that Lauren finished faster than Jo (an ordinal measurement), you also know *how many seconds* faster—you know the size of the *interval*. Ratio data is the same idea, except that it is measured on an absolute scale with a zero in it (so ratios are meaningful).

For example, if you know that Jo ran a race in 60 seconds and Lauren ran a race in 30 seconds, you could create a meaningful ratio—Lauren is twice as fast as Jo. The scale has a zero in it—zero seconds has intuitive meaning—so the measurements reflect ratio data. If you only know that Lauren finished 10 seconds faster than Jo, you know the interval—how much faster—but you can't compute the ratio—they might have been running for seconds, minutes or hours.

Interval and ratio data are sometimes called parametric data. They allow use of statistical techniques based on averages. If Jim were going to compare averages across groups identified by nominal variables—average race time by gym class, or average race time for the group that used the equipment (vs. that which did not), he must use interval or ratio data to measure physical fitness.

The type of data limits the choice of statistical techniques. So, if you can choose the type of data to collect, choose the type of number that provides the most information: ratio over interval, interval over ordinal, and ordinal over nominal. Why? The length of a board (a ratio measurement) also provides interval information (the difference in length between two boards), ordinal information (you could rank them by size), and nominal information (which is "short" and which is "long"). However, if you record only nominal information ("short" and "long"), the other information is lost.

Exercise #6

1. Identify the type of data in the following examples (ratio, interval, ordinal or nominal).

   a. The length of bars of steel _____

   b. One car is 10 miles ahead of another and 30 miles behind a third _____

   c. Personality types: introvert, extrovert _____

   d. Mortgage amount in dollars _____

   e. A rating "on a scale of 1 to 10" _____

   f. Birth order of brothers and sisters _____

   g. Finish order in a 10K race _____

   h. Time between finishers in a 10K race _____

   i. Time from start to finish in a 10K race _____

   j. Classification of times as "slow" or "fast" in a 10K race _____

2. Given a choice of all four types of numbers, which type provides the most information? The least? Why?

   _____

   _____

   _____

   _____

   _____

   _____

(*The answers to question 1 are on page 113.*)

HOW TO BUILD A DATA FILE

After collecting appropriate data, the next step is to organize your measurements into a *data file*. (A data file is sometimes called a *data set*.) Most statistical software can use a file similar to the one described in this section. Even if you are creating statistics by hand, you will still need to code and organize your data in a similar way.

Let's create a data file to solve Jim's problem (see Figure 1 on the next page). He is advancing a theory that physical fitness is a function of equipment. He defined *equipment* as a nominal variable, (*yes* or *no*), depending on whether an individual was enrolled in a gym class authorized to use the new line of equipment. He defined *physical fitness* as the finish order in a race. He also decided to collect several other measurements that might be useful.

Using standard terms, his data could be described like this:

- **VARIABLE**

    A specific measurement recorded for each individual. Jim's independent variable is *physical fitness*. Jim's dependent variables are *name*, *height*, *gender*, *equipment*, and *finish order*. *Name*, *equipment* and *gender* are nominal variables. *Height* is a ratio variable. *Finish order* is an ordinal variable.

- **CASE**

    A particular individual. Case #1 contains information about a student named *Judy*.

- **VALUE**

    The score defined by a particular case and a particular variable. Judy's value for the variable *height* is 68 inches (a ratio variable).

- **DATA FILE**

    A table listing values of variables by case. Depending on your needs, each column identifies a variable and each line identifies a case; or conversely, each column could identify a case and each line a variable.

Figure 1: How to Build a Data File

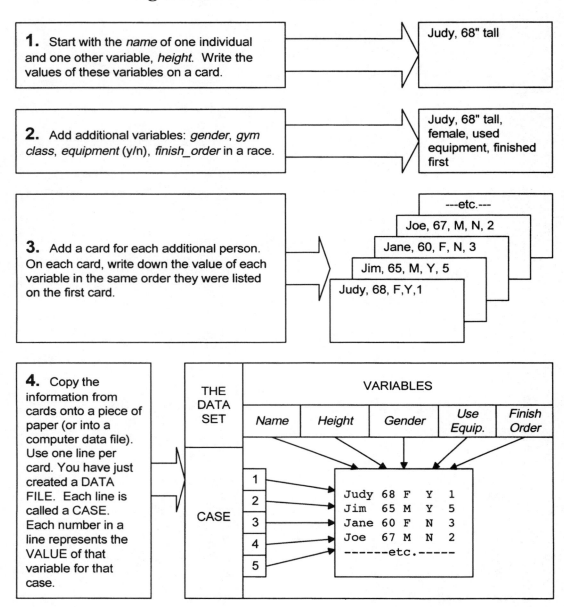

1. Start with the *name* of one individual and one other variable, *height*. Write the values of these variables on a card.

Judy, 68" tall

2. Add additional variables: *gender*, *gym class*, *equipment* (y/n), *finish_order* in a race.

Judy, 68" tall, female, used equipment, finished first

3. Add a card for each additional person. On each card, write down the value of each variable in the same order they were listed on the first card.

---etc.---
Joe, 67, M, N, 2
Jane, 60, F, N, 3
Jim, 65, M, Y, 5
Judy, 68, F,Y,1

4. Copy the information from cards onto a piece of paper (or into a computer data file). Use one line per card. You have just created a DATA FILE. Each line is called a CASE. Each number in a line represents the VALUE of that variable for that case.

THE DATA SET

CASE

	VARIABLES				
	Name	Height	Gender	Use Equip.	Finish Order
1	Judy	68	F	Y	1
2	Jim	65	M	Y	5
3	Jane	60	F	N	3
4	Joe	67	M	N	2
5	------etc.-----				

To create a data set, decide what variables to measure and record them in a table. For input, most statistical software accepts a text file organized like the table in step 4. Cards are used only to illustrate the thought process.

Exercise #7

1. Name three dependent variables that could be used as additional operational definitions of physical fitness. What kind of numbers are they (nominal, ordinal, etc.)?

    *   _____

    *   _____

    *   _____

2. On a separate sheet of paper, design a data set for a research problem at work or school. Use a spreadsheet format. For column headings, list the variables you would like to measure. Don't label the rows. Simply describe what constitutes a case for this project.

3. If you have time, review your analysis of the introductory stories in each of the following chapters. Design a data file for each. Define exactly what constitutes a case.

Summary

Measurements often represent different types of information, and the type of information limits the choice of statistical tools. Nominal measurements are names or categories. Ordinal measurements indicate ranking. Interval measurements tell the difference between measurements. Ratio measurements use an absolute scale with a zero in it. Data files organize the information like a spreadsheet. Each column represents a different variable, and each line represents a different case (or vice versa). An individual number in a data set represents the value of a particular variable for a particular case.

> . . . Jim created a data file using each student as a case. Each case contained measurements for several variables that represented different measures of physical fitness. Each case also included nominal variables that indicated the student's gym class, and whether that class used the equipment.

> The analyst opened the meeting: "Sorry I couldn't talk longer the other day, but this file looks like just what we need. Now, let's talk about what the school board wants to see." . . .

SECTION 4
Choose Cases to Study

Kim, the plant manager, was in conference with Raul, the company's labor negotiator, preparing for contract negotiations.

"I'd like to avoid a strike this year, if possible, Kim," Raul said, "Do you know what issues are the most important out there on the factory floor?"

Kim paused. He had some idea, of course, but he wanted to be more certain. Yet there were 2,500 employees out there. He knew what he wanted to measure, but there was simply no time to talk to everyone, and the new contract was too important to leave to guesswork.

What would you do?

CASES, SAMPLES AND POPULATIONS

Kim knows what his data file should look like, but he needs to fill it with data. Like most researchers, he will probably select a *sample* of *cases* from the *population*.

Cases

A case does not need to represent a person; it could represent many other things. There are as many definitions of case as there are research problems.

- It could be a production lot. Variables could be the *number of items* produced, *number of defectives*, *time of day* produced, etc.

- A case could be a particular calendar year. Variables could be *population*, *production of pig iron*, *miles of highway built*, etc.

- It could be a customer. Variables could be the *date of first purchase*, *date of last purchase*, *average amount purchased*, *zip code*, etc.

The research problem also determines how many variables and cases are needed. Specific problems can require only a few variables. For example, your bank account is like a data file with only two variables (*account number* and *transaction amount*) and many cases (one for each check or deposit). Complex problems often require many more variables. For example, your annual tax returns could be summarized in a data set with many variables—one for every line on every tax form you could potentially file—but few cases (only one for every year you file a tax return). In general, the broader the question, the more information you will need to answer it.

The definition of a case, who or what you will study, is determined by the definition of the population, which is determined by the research problem. In the introductory story for this section, Kim's research problem dealt with employee opinions at this plant. So, his population was defined as employees who worked at the plant, and each case represented a selected member of the population: an employee who worked at the plant. In the introductory story for Section 3, Jim defined the population as students enrolled in physical education at the school, and each case represented a selected member of that population: a student. In the introductory story to Section 2, Toni's population was employees who had left the company, and she organized her data using one case per employee.

Samples and Populations

The set of all possible cases is called a *population*. A group selected from the population is called a *sample*. The population of North Americans with bachelor's degrees is everyone on the continent who qualifies. A sample of such individuals might be only 50 people in a market research study. It is seldom practical to study an entire population, so most researchers use samples most of the time.

How big should a sample be? In general, the bigger the sample, the more accurate the results. Also, when the number of cases in the sample is above 30, some tools become appropriate that can't be used with smaller samples. You may need a pilot study to see how large the actual sample should be. At other times, you may wish to consult a statistics text for ways to estimate the required sample size using assumptions and formulas.

Exercise #8

Describe three different populations from your school or work. For each, identify three or four problems that research might help solve. For each problem, what is the definition of the population? Of a case?

1. _____

_____

_____

2. _____

_____

_____

3. _____

_____

_____

RANDOM SAMPLES

For a sample to represent a population accurately, every member of the population must have an equal chance of being selected. This produces a *random sample*.

Avoid sample bias. Samples that are not random can produce *biased* results. If Kim handed out questionnaires in front of the manufacturing plant, would he obtain a random sample? If the questionnaires were handed out at the start of first shift, they would not represent evening workers. If the lab were on the other side of the plant, those workers might be excluded. If most employees used other gates, the survey might represent only a few departments.

How to select a random sample. A better approach would be to obtain a master list of all employees, and use a table of random numbers to select the individuals to survey. (Almost all statistics books include tables of random numbers.) Here is one method. Let's say there were 2,500 employees on the list and Kim needed to select a random sample of 100. (This process is very simple, so don't try to make it more difficult than it is.)

▶ *Locate a table of random numbers and pick an arbitrary starting place.* Kim should close his eyes and let his finger fall anywhere on the page. To keep from always starting at the same place, he should pick the last two or three digits of whatever number is just above his fingertip, and move his finger that many more numbers down the page (or across the page, or diagonally, or whatever). If that position is off the page, he should just move to the next page of random numbers. At the end of the table, he should simply move back to the beginning and continue.

▶ *Keep the number, or throw it away.* Kim is now pointing to his starting place. Since he is trying to select a sample of 100 from a four-digit population (2,500 individuals), he should read the last four digits of the number above his fingertip. If it is between 1 and 2,500, he should use it—write it down on a list—and skip to the next number. If the last four digits are not between 1 and 2,500, he should throw it away—ignore it and skip to the next number. For example, if the random number from the table were 19,012, he would throw it away—not use it—because the last four digits (9,012) were not between 1 and 2,500. On the other hand, if the random number were 21,576, he would keep it—write it down on the list—because the last four digits (1,576) are between 1 and 2,500; they identify an employee to include in the sample: number 1,576 on the list of 2,500.

▶ *Repeat until the sample definition is complete.* Kim should continue moving down the column of numbers, keeping or using each, until his list of keepers is as big as the sample he intends to select (100 individuals in this example). If he runs out of random numbers because the table is too short, he could find a bigger table, switch to the *first* five digits, etc.

Exercise #9

1. Television stations often report polls of "people who chose to respond." Is that a random sample? Why or why not? _____

_____

_____

2. What are the implications for the results of such polls? _____

_____

_____

3. Describe how you would select a random sample for each of the populations you identified in the last exercise. _____

_____

_____

4. Describe another sampling process that might produce a biased sample of each population. _____

_____

_____

(The answers are on page 113.)

EXPERIMENTAL DESIGN

Besides introducing error through the way he selects his sample, Kim could also introduce error by the way he conducts the study itself. His survey should not signal the preferred or expected answers by using leading questions. He should be careful with socially sensitive questions that people might not answer accurately. If necessary, he should qualify the results when individuals who respond appear to be unlike those who choose not to respond. Kim should also be sensitive to results that could be distorted because individuals have been surveyed or tested before. Finally, he should remember that just the act of studying a problem often has an effect that can distort results.

Experimental design can help solve some of these problems. Think about measuring the impact of a training seminar in presentation skills. A one-shot (nonexperimental) design would only test performance of one group after the seminar. A two-group, before-and-after design involving a control group and an experimental group would be better. First, randomly assign employees to two groups and test both *before* the seminar, perhaps by having their supervisors rate their presentation skills. Then, while one group attends the seminar on presentation skills, send the control group to a seminar on an unrelated topic, such as workplace safety. After the seminar, test both groups again. By comparing the experimental group to the control group, both before and after, this experimental design would isolate the effect of the presentation skills training.

Experimental designs can be very complex. However, even in a simple study, try to use before-and-after testing, control groups, random sampling, random assignment to experimental and control groups, and other principles of good design. Be sure your data file includes variables to indicate the different groups (or whatever) that your experimental design requires.

Exercise #10

1. Define the following:

   a. Population _____

   b. Sample _____

   c. Random sample _____

   d. Biased sample _____

   e. Experimental design _____

2. For a research problem you are now facing, discuss possible sources of error and ways to eliminate them. Also sketch out an experimental design you could use to test the problem.

Summary

After designing a data file, you must fill it with data. Since testing an entire population is usually difficult, most of the time you will use a sample. The sample must be large enough to properly represent the population. To generalize from sample results to population conclusions, you must use a random sample: one in which every member of the population has an equal opportunity to be selected. Good experimental designs can help avoid, minimize and control sources of error.

> . . . Kim replied, "We've still got some time before contract negotiations start. I'd like to run a quick employee survey, using a sample of 100 workers. I'll use employment lists and a table of random numbers to select a true random sample. The survey will be anonymous, and before we hand it out, we'll test it on a small group of employees, just to be sure it makes sense and doesn't have any leading questions. I think it will help us tailor our contract proposals to what the workers really want." . . .

P   A   R   T

II

The Two Keys to Understanding Statistics

SECTION 5

Summarizing Numbers

Jennifer was meeting with her production managers. "Why are we still behind schedule?" she asked.

"It's all the prep shop's fault," Jordan replied. "Raw material stock is never the right length. We always have to adjust the machinery to fit different sizes."

"Not true!" the prep shop manager exclaimed.

The argument continued for some time and Jennifer finally gave up. "Both of you meet me in my office this afternoon. Bring the shop records. Let's find out who is right."

That afternoon both managers appeared along with three four-drawer file cabinets. Jennifer studied a few of the files and threw her hands up in desperation. "This 'information,' if you can call it that, is all over the place. We can't use this. What do you suggest we do?"

What would you do?

WHAT IS A SUMMARIZING NUMBER?

Data sets, also called data files, contain many numbers, sometimes millions of them. Even for a single variable, all of those individual cases are difficult to comprehend. So, it is often necessary to condense information from *many* numbers into *one* meaningful number. This number is called a *summarizing number*.

A summarizing number is a single number whose value is determined by all of the numbers for a particular variable in a particular data set. This is a very important section because the idea of a summarizing number is one of two key concepts in statistics. (Distributions, discussed in the next section, is the other key concept.) Although the phrase sounds intimidating, the idea behind summarizing numbers is very simple. You used one in elementary school:

- **INPUT:** A list (or data set) recording the height of each person in a group.

- **PROCESS:** Total the numbers. Divide that result by the number of people.

- **OUTPUT:** The average height for the group.

The output (the average) is a summarizing number. There is only one value for any given variable and data set. (In a group, there is only *one average* height, although there are *many individual* heights.) A summarizing number condenses the information from the entire group of numbers, and each number in the group has an impact on the summarizing number's value.

You are also familiar with several other summarizing numbers: maximum, minimum, and number of values (or cases). Since these functions are so widely used, they are included in most spreadsheet software (@AVG, @MAX, @MIN, @COUNT, etc.). To use these functions, you must specify a range of numbers that acts as a data set. The function returns a number that acts as a summarizing number. It *condenses* information from the range into a single number.

Research studies use many different summarizing numbers (at least one per variable, usually more). Figure 2 on page 35 uses five different summarizing numbers that could apply to a single variable. Figure 3 on page 38 describes seven more.

Although unchanging for a *particular* set of data, for *different* sets of data a summarizing number will usually vary across a wide universe of possibilities. For example, a group average of anything usually changes (somewhat) whenever any different group is chosen. When dealing with this universe of possible values, statisticians call a summarizing number from one of them the value of a *random variable* (the one that applies to that particular data set). Practically, when dealing with samples—as most business research does—the terms *statistic*, *the value of a particular random variable*, and what this book calls a *summarizing number* are essentially interchangeable.

Exercise #11

1. Define the expression *summarizing number*. _____

_____

_____

_____

_____

_____

2. Why are summarizing numbers used? _____

_____

_____

_____

_____

_____

MEASURES OF CENTRAL TENDENCY

The summarizing numbers *mean*, *median* and *mode* all measure *central tendency*: a characteristic of data to cluster around some value (which is often somewhere near the middle of all the others). For example, how would you describe the height of a group of individuals? You might answer in three different ways: "They average six feet tall," or "Half are over six feet one inch," or "The most frequent height is five feet, eleven inches." Statisticians have specific names for the summarizing numbers you just used: the mean, the median and the mode.

► **MEAN:** The total of all values, divided by the number of cases. It can only be used with interval or ratio numbers. (This is commonly called the average.)

► **MEDIAN:** The value such that 50% of cases are equal to or below it and 50% are equal to or above it. It can be used with interval, ratio and ordinal numbers. If no number splits the ordered data set into two equally sized groups, the median is halfway between the two closest candidates.

► **MODE:** The most frequent value. It is appropriate for all types of numbers (nominal, ordinal, interval and ratio) where at least one value occurs more frequently than most of the others.

The mean is usually the best indicator of central tendency, but not always. For nominal numbers, it is meaningless; use the mode instead. For ordinal numbers, use the median. Even when the mean can be used, the median is a better estimate if a few extreme values distort the mean.

Figure 2: Central Tendency Calculations

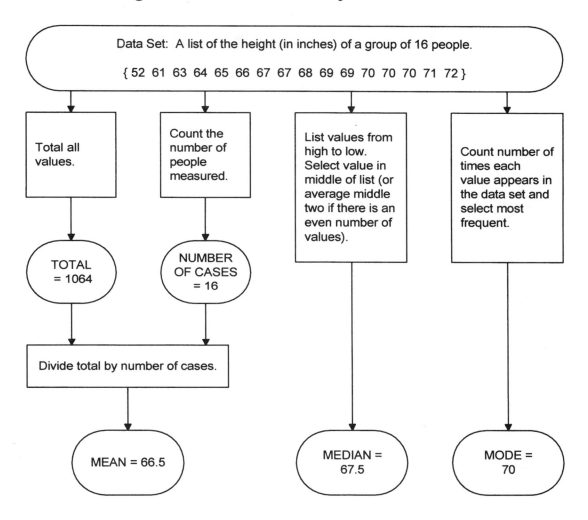

Data Set: A list of the height (in inches) of a group of 16 people.

{ 52 61 63 64 65 66 67 67 68 69 69 70 70 70 71 72 }

Total all values.

Count the number of people measured.

List values from high to low. Select value in middle of list (or average middle two if there is an even number of values).

Count number of times each value appears in the data set and select most frequent.

TOTAL = 1064

NUMBER OF CASES = 16

Divide total by number of cases.

MEAN = 66.5

MEDIAN = 67.5

MODE = 70

Summarizing numbers condense information from many individual measurements into a single number. The summarizing numbers used in this example (mean, median and mode) are called measures of central tendency.

Exercise #12

1. Name the three measures of central tendency. What types of data (nominal, ordinal, interval or ratio) are appropriate for each? Give an example of appropriate data.

	Name	**Data Types**	**Example of Data**
a.	_____	_____	_____
b.	_____	_____	_____
c.	_____	_____	_____

2. From the following lists of data, calculate the indicated summarizing numbers. Assume ratio data, if possible.

Data Set	Total	# of Cases	Mean	Median	Mode	Max	Min
10,10,10,15,15,60							
2,3,5,7,7,8,10							
2,11,11,13,28							
red, red, blue							
2,2,4,5,102							

3. Under what conditions is the median a superior measure to the mean?

_____

_____

(The answers to question 2 are on page 113.)

MEASURES OF VARIABILITY

It is very useful to know how individual scores for a variable *vary* around the mean.

The idea of variability sounds complex, but it is really very simple. If you were planning a trip to a tropical island, it would be easy to pack. Daily temperatures wouldn't vary over 10 degrees from the annual average. However, if the trip were to the American Midwest, the same mean annual temperature would be the result of blizzards and heat waves. The daily temperatures there are much more widely dispersed around the mean. They are much more variable.

Variability is also a useful measure of quality. A reduction in variability means an increase in quality, because the product is consistently closer to the design. If the blueprint calls for a 0.5 inch hole, a plus or minus 0.01 inch process is of higher quality than one where holes average plus or minus 0.03 inch—the holes are less variable.

To measure this *dispersion* of scores around the mean, statisticians use two summarizing numbers: the *standard deviation,* or its close cousin, the *variance*. The standard deviation is a very similar concept to the *average distance from the mean* for a typical value. The variance is simply the square of the standard deviation. A *smaller* standard deviation (or variance) means scores are generally *closer* to the mean. A *larger* standard deviation (or variance) means scores are generally *farther* from the mean—they are more spread out, more variable. The standard deviation is useful because it uses the same units as the original measurement (for example, the "standard deviation is four inches"). The variance is used in advanced statistics.

Compared to other summarizing numbers, the standard deviation and the variance are the best ways to measure variability. The range (the difference between the largest and smallest values) also indicates variability, but it can be distorted by only one extreme value. Another solution is to measure the absolute difference between the mean and each value, then calculate the average of those differences. An example of this *mean absolute deviation* (MAD) is in Figures 3 and 4. It is a better measure of variability than the range, but it is hard to work with mathematically, so statisticians do not use it.

EXAMPLES OF MEASURES OF DISPERSION AHEAD

Figure 3: Dispersion Calculations

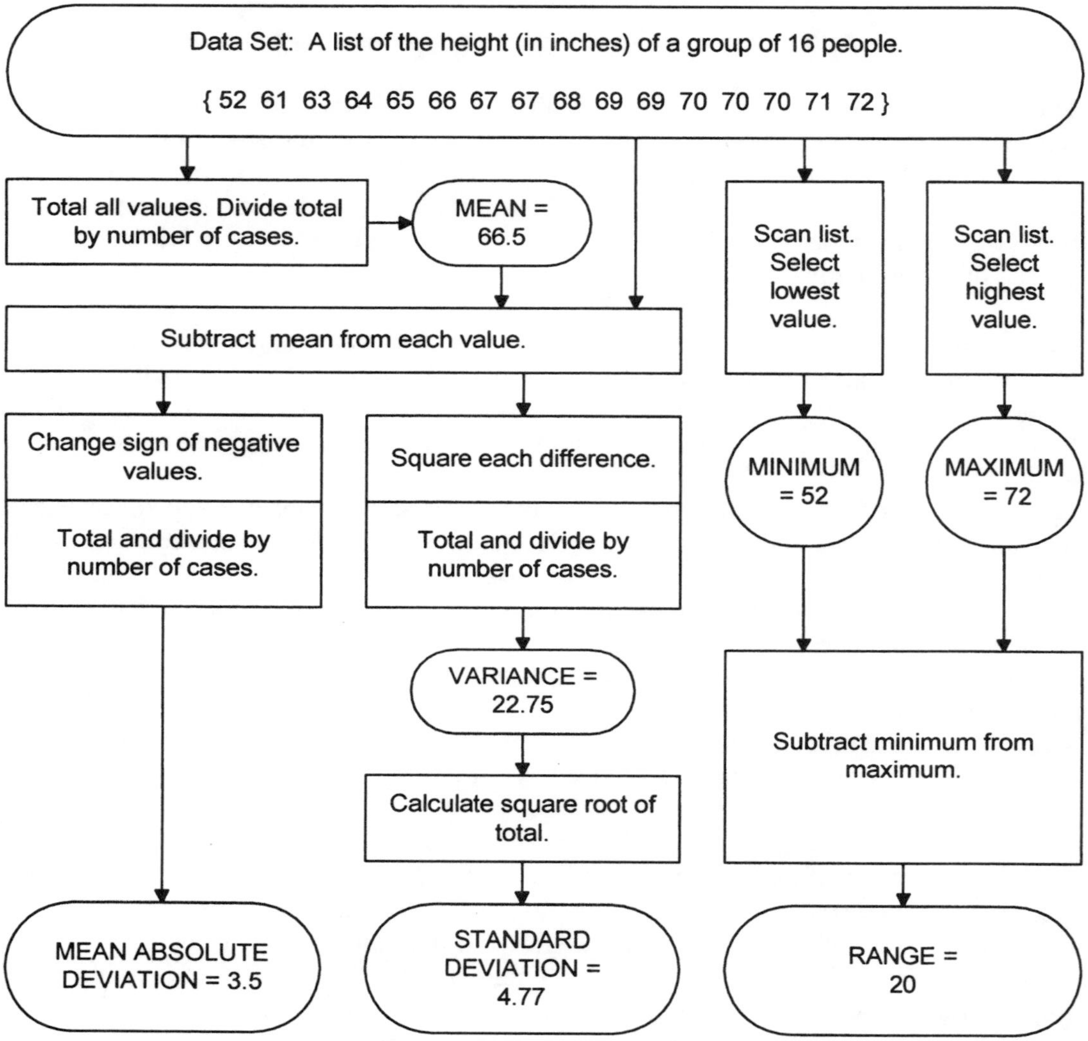

Data Set: A list of the height (in inches) of a group of 16 people.

{ 52 61 63 64 65 66 67 67 68 69 69 70 70 70 71 72 }

Total all values. Divide total by number of cases.

MEAN = 66.5

Scan list. Select lowest value.

Scan list. Select highest value.

Subtract mean from each value.

Change sign of negative values.

Total and divide by number of cases.

Square each difference.

Total and divide by number of cases.

MINIMUM = 52

MAXIMUM = 72

VARIANCE = 22.75

Calculate square root of total.

Subtract minimum from maximum.

MEAN ABSOLUTE DEVIATION = 3.5

STANDARD DEVIATION = 4.77

RANGE = 20

These summarizing numbers (variance, standard deviation, mean absolute deviation and range) are called measures of dispersion because they suggest how widely individual scores are scattered (dispersed) about the mean. This chart shows the population formula; for samples, the formula is slightly different.

Figure 4: What is a Standard Deviation?

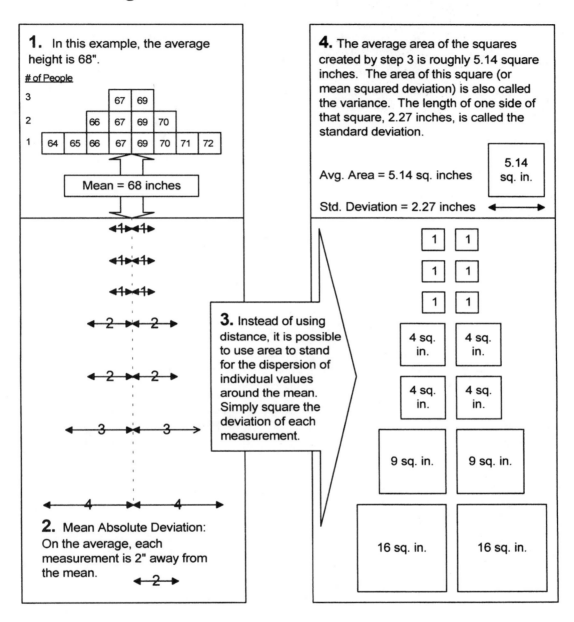

1. In this example, the average height is 68".

# of People

3 | 67 | 69
2 | 66 | 67 | 69 | 70
1 | 64 | 65 | 66 | 67 | 69 | 70 | 71 | 72

Mean = 68 inches

2. Mean Absolute Deviation: On the average, each measurement is 2" away from the mean.

3. Instead of using distance, it is possible to use area to stand for the dispersion of individual values around the mean. Simply square the deviation of each measurement.

4. The average area of the squares created by step 3 is roughly 5.14 square inches. The area of this square (or mean squared deviation) is also called the variance. The length of one side of that square, 2.27 inches, is called the standard deviation.

Avg. Area = 5.14 sq. inches

5.14 sq. in.

Std. Deviation = 2.27 inches

The standard deviation is very similar to mean absolute deviation. The MAD measures the average distance to the mean. The variance and standard deviation use the corresponding area measurement. Both measure expected variation, but the area measurements work better mathematically.

As Figure 4 shows, the variance carries a very similar intuitive meaning to the MAD. The MAD can be visualized as the length of a line that corresponds to the average distance of a typical value from the mean. The variance can be visualized as the *average area of a square* (whose side is closely related to the distance from the mean to the typical value). The length of a side of this representative square (the square root of the variance) is called the *standard deviation*. As variability goes up, the MAD also goes up (the line gets longer). So does the variance (the area increases). So does the standard deviation (the side of the square gets longer).

Exercise #13

1. Name four measures of deviation from the mean. Which is best? Why?

    a. _____

    b. _____

    c. _____

    d. _____

2. From the following data sets, use a calculator and Figure 3 to calculate the values of the following summarizing numbers. (Use the formula for the variance and standard deviation of a population, which is the one shown in Figure 3.)

Data Set	Variance	Standard Deviation	Range
50, 52, 48, 54, 46			
50, 45, 55, 40, 60			
50, 25, 75, 25, 75			
50, 0, 100, 25, 75			
50, 52, 48, 52, 48			

(The answers to question 2 are on page 113.)

FORMULAS AND NOTATION

Statisticians use common symbols to make working with statistics easier. After all, calculating a summarizing number requires reading the value of a particular variable for *every case* in the entire data set. Formulas written the usual way would be very long, and a different one would have to be written for every possible size of data set.

To avoid these complications, statisticians use certain conventions to write statistical formulas.

Σ The summation symbol is the uppercase Greek letter sigma. It means calculate the value of the formula to the right after substituting the first value (of a given variable in the data set) for x. Repeat, using the second value, the third, etc., until you reach the last case. Then, total the results of all the calculations and use that total for the answer.

x x stands for the value of the variable for a single case. As you repeat the formula after a summation sign, x first takes on the value of the variable for the first case. The next pass, it takes on the value of the variable for the second case, etc.

n n stands for the number of cases in a data set. Specifically, a lowercase n stands for the number of cases in a sample. An uppercase N stands for the number of cases in the entire population.

A random variable for a population is called a *parameter* and is indicated by a lowercase Greek letter. A summarizing number for a sample is called a *statistic*, and is usually indicated by a lowercase English letter. Here are the formulas for the three summarizing numbers we have just discussed.

The Mean

The lowercase Greek letter mu (μ, pronounced "mew") is used to represent a mean for an entire population.

$$\mu = \frac{\Sigma(x)}{N}$$

For a sample, the symbol is x bar. As you can see, the formula is the same.

$$\bar{x} = \frac{\Sigma(x)}{n}$$

The Standard Deviation

The symbol for the standard deviation is the lowercase Greek letter sigma (σ) if it refers to an entire population.

$$\sigma = \sqrt{\frac{\Sigma(x - \mu)^2}{N}}$$

For a sample, the lowercase English letter s is used.

$$s = \sqrt{\frac{\Sigma(x - \bar{x})^2}{(n - 1)}}$$

Notice that for a *sample* standard deviation, the number of cases is reduced by one. This minor change produces a better estimate when the sample mean is used to estimate the true (population) mean. Don't be intimidated by the formula. It is the same process indicated in the flowchart you just studied in Figure 3. If it is unclear, spend a couple of moments with the flowchart, tracing out the formula. (Spreadsheet software usually includes functions to calculate variance and standard deviations for both populations and samples.)

The Variance

The variance is simply the square of the standard deviation.

Population:

$$\sigma^2 = \frac{\Sigma(x - \mu)^2}{N}$$

Sample:

$$s^2 = \frac{\Sigma(x - \bar{x})^2}{(n - 1)}$$

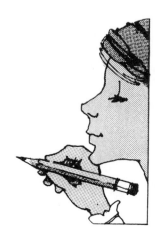

Exercise #14

1. Define these common symbols:

Σ	
σ	
μ	
$\overline{x}$	
s	
x	
s^2	
σ^2	

2. What is the difference between σ and s? Between μ and $\overline{x}$? _____

_____

_____

_____

_____

_____

(The answers to question 1 are on page 113.)

Summary

A summarizing number condenses the information from many measurements of a variable into a single number. Common summarizing numbers include the number of cases, maximum, minimum, mean, median, and mode. The mean, median and mode are also called measures of central tendency. The mean is usually the best measure of central tendency. Other summarizing numbers measure the dispersion of values around the mean. The variance and standard deviation are the best measures of dispersion. Statistical formulas use certain common symbols. Summation symbols (Σ) indicate you should repeat the formula to the right for the specified elements of the data set (usually every element) and sum the results. Summarizing numbers calculated for a population are called parameters and are indicated by lowercase Greek letters. Summarizing numbers calculated for a sample of the population are called statistics and are indicated by lowercase English letters. The symbol for the mean is either μ or x bar. The symbol for the standard deviation is either σ or s.

> . . . Jennifer recovered quickly after reviewing the data provided by her production managers. "Take all this stuff out of here. One week from today, bring me two graphs. The first graph should show the mean length of raw materials leaving the prep shop by month for the last year. The second should show the standard deviation for the same time period. Two weeks from today, let's meet and decide how we can reduce the standard deviation to 75% of its current level, whatever that is, by the end of next month." . . .

SECTION 6

Distributions

Randy was listening to Kelly, the Personnel Manager, discuss results from a training seminar. "We gave a test before and after the seminar," Kelly said. "Overall, it showed improvement. I recommend we train everyone."

"To do that," replied Randy, "we've got to sell senior managers on the benefits. They really like graphs. Can you somehow give me a picture of the results?"

"I have an idea," Kelly answered. "I'll be back this afternoon."

If you were Kelly, what would you do?

WHAT IS A DISTRIBUTION?

In almost every field of study, a few fundamental concepts provide the foundation for almost everything else. In statistics, there are two: *summarizing numbers* and *distributions*. A distribution is an association between different values of a variable and *another measurement*. This association can be described using a table, a graph or an equation. Every time you hear the word distribution, think of the two elements: the variable and its associated measurement.

When the associated measurement is the *number of cases*, the result is a *frequency distribution*. When presented as a graph, a frequency distribution is called a *histogram*. In Figure 5 are two examples. The first shows scores for a classroom test. The leftmost bar on this histogram says that one student scored 81 on the test; the next bar to the right says that one student scored 83 on the test; the next, that three students scored 84, two scored 85, etc., until the scores of the entire class, all 30 students, have been reflected on the graph. The second example shows the percentage of defective parts in each lot of a manufacturing process. In this second histogram, the leftmost bar on the chart says that one production lot had 0.5% defective; the next, that four production lots had 1% defective; the next, that 13 lots had 1.5% defective, etc.

Exercise #15

Sketch out a histogram that would be useful in your work or in an academic research project. Identify the variable. What is the associated measurement? What do you think the most frequent value would be for the variable? The least frequent? Can you estimate how the graph would be shaped? What would a change in shape tell you?

Number

of

Cases

Variable (specify) _____

Figure 5: Examples of Frequency Distributions

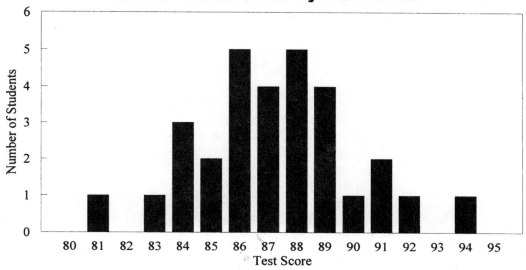

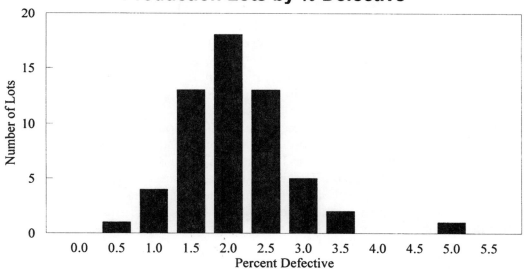

The Sand Pile Distribution

Probability distributions are an even more useful type of distribution than frequency distributions. Probability distributions are general-purpose tools that can be used as models of many different things.

Let's build one in your front yard. Imagine a truck dumping a load of sand in your driveway. Every grain of sand won't fall in the same place, will it? Instead, the truck will leave a big pile of sand. Imagine walking from your garage to the street, across the pile. At first, you would not step on any sand at all. Then you would step on a few grains, then it would be ankle deep, then knee deep, and finally you would reach the top of the pile. Continuing your walk, it would again be knee deep, then ankle deep, then just a few grains, and then no sand at all. Finally, you would reach the street.

You just walked across a distribution. The variable is the *distance from the edge of the street*. The measurement associated with it is the *amount of sand*. With a tape measure, shovel, scales and a bucket, you could create a table that described the sand pile. It would be easy to identify the part of the sand pile that was between, say, four and six feet from the street. You could shovel the sand that was between those measurements into the bucket and weigh it. By repeating the process for other parts of the sand pile, you could eventually create a table listing the amount of sand in all the different parts of the pile as a function of the distance from the street.

You could also show the same information as a graph. (The graph and the table are both in Figure 6 on the next page.) Let's say a bulldozer rumbles down your driveway and pushes away half the sand pile. (The sand is wet, so the remaining half is undisturbed.) The result is as if someone cut vertically through the middle of the pile with a giant knife and made half of it disappear. A picture of the flat side where the bulldozer passed would look just like the graph in Figure 6.

Can you see how the top of the sand traces a curve? Can you see how the distribution of the *area underneath the curve* corresponds to the actual distribution of the *sand on the driveway*? So, examining the area underneath this curve is the same as examining the actual distribution of the sand. The distribution of the area under the curve is an appropriate model for the physical distribution of the sand on the driveway. Probability distributions use the area under a curve in the same way. (Technically, discrete probability distributions are not continuous curves like this. The distinction is not important for the purposes of this book.)

Figure 6: The Sand Pile Distribution

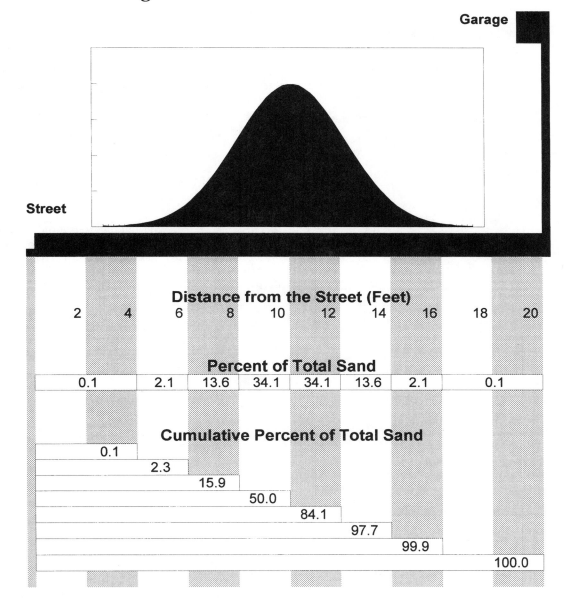

Garage

Street

Distance from the Street (Feet)

2	4	6	8	10	12	14	16	18	20

Percent of Total Sand

0.1	2.1	13.6	34.1	34.1	13.6	2.1	0.1

Cumulative Percent of Total Sand

0.1

2.3

15.9

50.0

84.1

97.7

99.9

100.0

(Sum of parts may not exactly agree with totals due to rounding.)

(To make statistics easier to understand, all the tables in this book use rounded values, so the sum of the parts may not exactly agree with the totals shown. If your project requires more precise values, be sure to consult the more detailed tables in any college-level statistics text.)

Exercise #16

1. Use either the table or the graph of the Sand Pile Distribution in Figure 6 to answer the following questions:

    a. What percentage of the sand pile was closer than 8 feet to the street? _____

    b. Farther than 14 feet from the street? _____

    c. Between 8 and 12 feet from the street? _____

    d. Greater than 10 feet or less than 6 feet from the street? _____

    e. Less than 14 feet from the street? _____

2. List three other examples of distributions you encounter in everyday life. For each, identify the variable and its associated measurement. Describe the rough shape of each. Example: A company advertises job openings in the Sunday paper. Weekly job applicants (associated measurement) are distributed by day of the week (variable). The distribution peaks on Monday and Tuesday, and trails off the rest of the week.

    a. _____

    _____

    _____

    b. _____

    _____

    _____

    c. _____

    _____

    _____

(The answers to question 1 are on page 114.)

There are many different probability distributions. Different distributions are needed because various situations require specific models.

Summary

A distribution is an association between different values of a variable and an associated measurement. A frequency distribution, or histogram, relates different values of a variable to the number of cases that occurred for each. A probability distribution relates different values of a variable to the area under a curve. The areas under certain curves can be models for many different real-world distributions.

> *. . . Kelly returned that afternoon with two distributions. One showed the number of managers across the range of scores for the pretest. The second showed the same information for the posttest. Randy smiled, "This is exactly what we need to justify the seminar. It really shows the difference, doesn't it?. . ."*

III

How to Turn Measurements into Answers

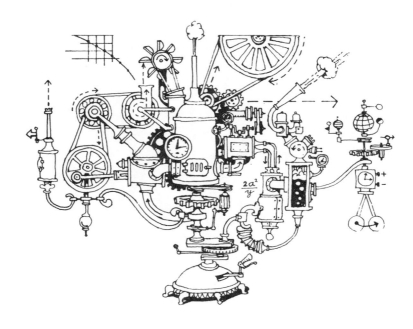

SECTION 7

Frequencies and Normal Curves

Problems with the prep shop had diminished, but they were still cutting a few bars to the wrong length.

"OK," said Jennifer, "the length of bars from the prep shop has a standard deviation of 1 centimeter. We're producing 10,000 bars per week. Why don't you just build storage for all of the bars over or under 2 centimeters from the specifications? Then, scrap or rework them together, say once per week."

"How many would that be?" the supervisor asked. "You figure it out," Jennifer replied.

What if you had to figure it out?

FREQUENCY ESTIMATES

Requirements:

- Interval or ratio data

- Frequency distribution matches normal curve

Purpose: Given value(s) of the variable, estimate number of cases

Distribution: Standard normal

THE STANDARD NORMAL DISTRIBUTION

The Sand Pile Distribution in the last section is really a *normal distribution*. The area underneath it is a good model for the distribution of many things besides sand. For example, instead of *distance from the street* you could substitute other variables: *height*, *weight*, *school test scores*, *percent defective parts* in a manufacturing process, etc. For these kinds of variables, it is likely the normal distribution would be a good model of the relative frequency of different scores.

Since the normal curve has so many different uses, it doesn't make sense to label it in units that apply to only one situation (such as buckets of sand, test scores or inches of height). Instead, we use the standard deviation, which is a measure that is common to all such uses.

Figure 7 shows a *standard* normal distribution because it has a mean of zero and a standard deviation of one. (Any normal distribution can be defined with only two numbers: the mean, and the standard deviation.) It is also known as the *z-distribution*, and values in it are called *z-scores*. Z-scores are composed of a number whose sign (+/−) indicates it is higher than (or lower than) the mean, and whose numerical part indicates how many standard deviations lie between that particular point and the mean.

Compare the picture of the Sand Pile Distribution in Figure 6 to the picture of the standard normal distribution (see Figure 7). (The standard deviation for the sand pile was two feet.) Can you see how 8 feet from the street corresponds to −1 standard deviation from the mean? How 16 feet from the street corresponds to +3 standard deviations from the mean? Do you see how the amount of sand closer to the street than 8 feet corresponds to the area under the curve that is left of −1 standard deviations? This is an important conceptual skill. Do not continue reading until you can switch back and forth mentally between *feet from the street* (or any other variable) and *standard deviations from the mean*.

Figure 7: Areas of the Normal Distribution

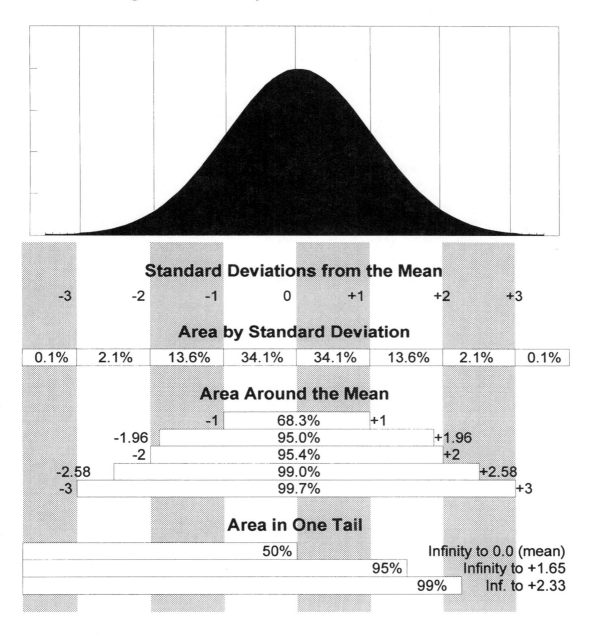

(Sum of parts may not exactly agree with totals due to rounding.)

RECOGNIZING NORMAL DISTRIBUTIONS

To use normal distributions, you must recognize them when they appear. The variable must contain interval or ratio data. The frequency distribution should be symmetric and bell-shaped like those in Figures 5, 6 and 7. The percent of total cases within 1, 2 and 3 standard deviations of the mean should be similar to corresponding areas under the standard normal curve. When these conditions are satisfied, you may use the standard normal probability distribution as a model for the frequency distribution of the variable. You can also conclude that the variable is normally distributed.

Of course, instead of a picture, most statistics texts provide much more detailed and precise information about the z-distribution by using tables. In a table, the same information in Figure 7 would look like this:

STANDARD NORMAL TABLE	
Z-score	**Cumulative Area Under the Curve**
–3.00	0.1%
–2.58	0.5%
–2.00	2.3%
–1.96	2.5%
–1.65	5.0%
–1.00	15.9%
–0.00	50.0%
1.00	84.1%
1.65	95.0%
1.96	97.5%
2.00	97.7%
2.33	99.0%
2.58	99.5%
3.00	99.9%

Remember, the tables in this book have been simplified. If your project requires more precise estimates of these values, consult a more advanced text. Can you trace the numbers from Figure 7 to this chart?

A TWO-STEP APPLICATION

When the normal distribution is a good model for the frequency distribution of a variable, it can answer very useful questions. These questions look like this: "What percent of total cases lie beyond (or between, or ahead of) some value of the variable?" For example:

- Engineers and industrial designers create autos, clothes, furniture, etc. to comfortably fit a particular size of individual. What percentage of the population would find these products uncomfortable?

- College admissions officers require a certain minimum performance on admission tests. What percentage of the student population will qualify?

- Managers monitor the percent defective for different lots in a production process. What percentage of all lots will exceed a certain percent defective?

- Questions on surveys can be analyzed with means and standard deviations. Given these measures, what percentage of the population is associated with various responses?

You may have recognized these as the same kind of problems you answered using the Sand Pile Distribution in the last section. Now we will turn this process into a recipe with two steps:

The Two-Step Recipe

STEP 1. Translate Critical Values into Z-scores

STEP 2. Look Up Critical Areas and State the Conclusion

Exercise #17

Identify three problems at work or school that could be analyzed using the normal curve and frequency analysis.

1. _____

2. _____

3. _____

Step 1: Translate Critical Values into Z-scores

Critical values are always identified in the problem. The question "How much sand is between 8 feet and 10 feet from the street?" has two critical values: 8 feet and 10 feet. The question "How much of the general population is taller than 5'6", has one critical value: 5'6". These critical values must be translated into z-scores, so you can look them up in a standard normal table.

To convert a measurement into a z-score, subtract the mean from that value, then divide by the standard deviation. A value of 15 taken from a variable with a mean of 20 and a standard deviation of 5 produces a z-score of –1. So, a score of 15 is one standard deviation away from the mean (to the left). A value of 25 is +1 standard deviation away from the mean (to the right).

$$z = (x - \bar{x}) / s$$
$$z = (15 - 20) / 5$$
$$z = -1$$

$$z = (x - \bar{x}) / s$$
$$z = (25 - 20) / 5$$
$$z = 1$$

Exercise #18

1. Calculate z-scores for the following scores. The first has been done for you.

Score x	Mean $\bar{x}$	Standard Deviation s	Z-Score $z = (x - \bar{x}) / s$
90	100	10	–1
12	10	1	
40	25	5	
95	100	10	
30	45	10	
425	500	25	

2. Circle z-scores in this list that are less than the mean: –1.5, 2, 3, –3, 1.5, –2.4.

(The answers are on page 114.)

The process also works in the other direction. The original measurement of a variable is the number of standard deviations away from the mean times the standard deviation, plus the mean. If the original distribution had a mean of 20 and a standard deviation of 5, then a z-score of -3 equals an original measurement of 5. A z-score of 2 equals an original measurement of 30.

$$x = (z * s) + \bar{x}$$
$$x = (-3 * 5) + 20$$
$$x = 5$$

$$x = (z * s) + \bar{x}$$
$$x = (2 * 5) + 20$$
$$x = 30$$

Exercise #19

1. Calculate original scores for the following z-scores. The first has been done for you.

Z-score	Standard Deviation	Mean	Original Score
z	s	$\overline{x}$	$x = (z*s) + \overline{x}$
−1.5	10	90	75
+2	1	15	
−3	5	35	
+2.5	10	65	
−0.5	2	84	
+1	5	20	

2. Given a mean of 100, circle the scores that would produce negative z-scores: 90, 85, 110, 124, 76, 99, 112.

(The answers are on page 114.)

Step 2: Look Up Critical Areas and State the Conclusion

Next, draw a mental picture of the distribution, cut it into pieces, and decide which piece will answer your question. Consider the question: "How much of the population exceeds 6'0" in height?" Mentally draw a picture of the distribution. Move right (from the center), and draw a vertical line through the critical point, 6'0". This separates the distribution into two parts: a small, right-hand piece (all individuals taller than 6'0"), and a large, left-hand piece (all individuals shorter than 6'0"). The question indicates you want the right-hand piece. It is the critical area. This is a *one-tailed analysis*—you only need the area contained in one tail of the distribution. (Do you see how this is exactly the same kind of question as "How much of the sand is greater than 14' from the street?")

Some questions require a *two-tailed analysis*. For example, "How much of the population is taller than 6'0" *or* shorter than 5'4"?" This question has two critical points. Vertical lines at these two points identify three areas: one representing heights greater than 6'0"; one representing heights less than 5'4"; and the middle range, which is everything else. The critical areas are the two extreme tails, the right-hand and left-hand pieces. (Do you see how this is exactly the same kind of question as "How much of the sand is closer than 6 feet to the street *or* farther than 14 feet from the street?")

Figure 8: One- and Two-Tailed Tests

One-Tailed Test

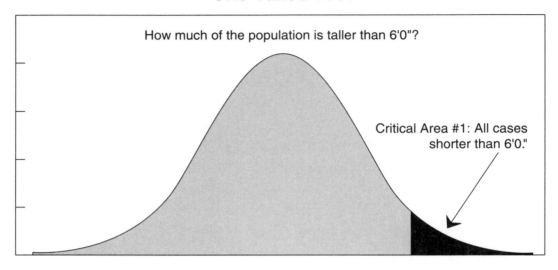

How much of the population is taller than 6'0"?

Critical Area #1: All cases shorter than 6'0."

Two-Tailed Test

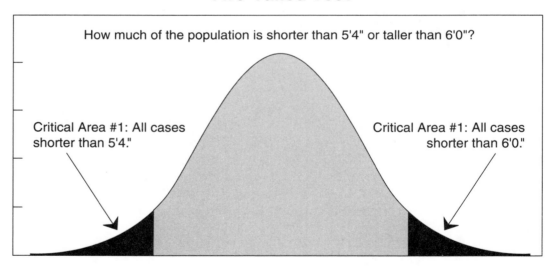

How much of the population is shorter than 5'4" or taller than 6'0"?

Critical Area #1: All cases shorter than 5'4."

Critical Area #1: All cases shorter than 6'0."

Exercise #20

You are running a manufacturing process where each lot yields an average of 3% defective parts. The percent defective statistic is normally distributed with a standard deviation of 0.5%. Decide whether the following questions call for a one- or a two-tailed test. To the right, sketch a picture of each distribution. Put a mark in the center for the mean. Draw vertical lines at the approximate locations of the critical points. Shade in the critical areas. Remember, for all practical purposes, a normal distribution is six standard deviations wide (three standard deviations on each side of the mean).

Percent Defective Parts	One Tail or Two?	Picture with Critical Area Shaded
Greater than 1.5%		
Less than 2.5%		
Between 1% and 2%		
Within 1% of the mean		

(The answers are on page 114.)

APPLYING THE TWO-STEP RECIPE

Let's trace a couple of examples through this process, using the situation from the previous exercise. From that exercise, you will recall that the average percent defective is 3% and the standard deviation of that figure is 0.5%.

Between what limits do 95% of production lots fall? This is a two-tailed test. Start with the standard normal chart (Figure 7). Look at the *Area Around the Mean* section. It says 95% of all cases fall within plus or minus 1.96 standard deviations from the mean. Therefore, the critical points are –1.96 and +1.96. So, 95% of all production lots have a defective percent between roughly 2% and 4%.

$$x_2 = (z * s) + \overline{x}$$
$$x_2 = (-1.96 * 0.5\%) + 3.0\%$$
$$x_2 = 2.02\%$$

$$x_1 = (z * s) + \overline{x}$$
$$x_1 = (1.96 * 0.5\%) + 3.0\%$$
$$x_1 = 3.98\%$$

How many lots exceed 4% defectives? This is a one-tailed test. Four percent represents a point 2 standard deviations from the mean. Consulting the standard normal chart (page 57), 2.1% + 0.1% = 2.2%. So, 2.2% of the area lies beyond 2 standard deviations from the mean. So 2.2% of all lots would be expected to exceed 4% defectives.

$$z = (x - \overline{x}) / s$$
$$z = (4.0\% - 3.0\%) / 0.5\%$$
$$z = 2.0 \quad standard\ deviations$$

Exercise #21

Work the following problems using the two-step method described previously. Consult the standard normal chart on page 57 to look up the areas. For each, assume the number of cases is normally distributed by the variable in question.

1. On a recent survey, customers rated your store an average of 4.0 with a standard deviation of 1.0. Anything greater than 5.0 represented negative feelings about the store. What percent of customers felt negatively about the store? _____

2. Every month customers return a mean of 2% of sales, with a standard deviation of 0.5%. What percent of the time would you expect to report returns of more than 3.0% of sales? _____

3. Average tenure among your employees is 5.5 years, with a standard deviation of 1 year. Of 100 individuals you hire today, how many would you expect to still be employed 4.5 years from now? _____

4. The average trip for your delivery trucks is 1,000 miles, with a standard deviation of 100 miles. Next year, you plan to make 1,000 trips. How many trips do you expect to make of less than 800 miles? _____

5. Monthly quality ratings for your department have averaged 85 with a standard deviation of 5. What percent of the time will the department probably score below 80? Above 90? _____

(The answers are on page 114.)

Summary

When the normal distribution is a good model for the frequency distribution of a variable, you can use an easy two-step process to estimate the number of cases above, below or between different values of that variable. Simply translate critical values into z-scores and look up associated probability in standard normal tables.

> . . . *"Well," the supervisor replied to Jennifer, "we know the number of bars is normally distributed by length, and we're after everything greater than 2 standard deviations from the mean—that's about 2.2% of the total. And 2.2% of 10,000 bars per week is 220; looks like we'll need a holding area with room for about that many bars." . . .*

SECTION 8

Contingency Tables

Avery was concerned about the number of new employees who were not considered promotable a year after they were hired. When a candidate accepted a position, the interviewer estimated probable promotability as high or low to facilitate planning for human resources. The assessment that really mattered (to the candidate) occurred 12 months later, when supervisors did the same thing.

"Maybe our interviewers don't know what supervisors value," Avery thought. "Or, perhaps we just aren't attracting promotable candidates. It would really be useful to know if the supervisors and the interviewers usually agreed. There must be some way to analyze this."

If you were Avery, what would you do?

CONTINGENCY TABLES

Requirements:

- Two or more nominal variables
- Two or more values of each variable
- Expected number of cases > 10 for every possible combination of values

Purpose: Discover relationships between concepts. Given one variable, can you predict the other?

Distribution: Chi-square

WHAT IS A CONTINGENCY TABLE?

Contingency tables (or *cross-tabulations*) allow you to explore the association between two or more concepts measured by nominal variables. Nominal variables are equivalent to a name (*gender*, *department number*, etc.). Contingency tables require that you simultaneously classify cases by at least two variables, with each variable having at least two values.

To solve Avery's problem, he needs to simultaneously classify employees by interviewer rating and by supervisor rating. That analysis might show a positive association (ratings are usually similar for interviewers and supervisors), a negative association (their ratings usually disagree), or perhaps no association at all (neither is useful in predicting the other).

Let's say Avery selected a random sample of 120 individuals. He cross-classified each case using interviewer and supervisor evaluations. The process and results might look like the chart in Figure 9. The table at the bottom of the chart is the contingency table.

A close look at the table shows how it was created. John (step 2) is one of the 42 individuals (step 4) who were rated high by both interviewer and supervisor. Sam (step 2) is one of the 60 individuals (step 4) who were rated low by both. You can create a contingency table by simply passing through the data set and making tally marks in the correct square.

Contingency tables can be very useful. For example:

- To test whether one geographical area is responsible for large orders, you could classify orders by *size* (large, small) and *geography* (area A, area B, etc.).

- To test whether sales of one product are associated with another, you could classify orders by two product types: *Product A ordered* (yes, no) and *Product B ordered* (yes, no).

When you hear the phrase *contingency table*, visualize a spreadsheet. The *rows* represent values of one variable. The *columns* represent values of the other variable. Each *cell* is an intersection of a row and a column.

IT'S EASY TO CREATE A CONTINGENCY TABLE ⟩

Figure 9: How to Create a Contingency Table

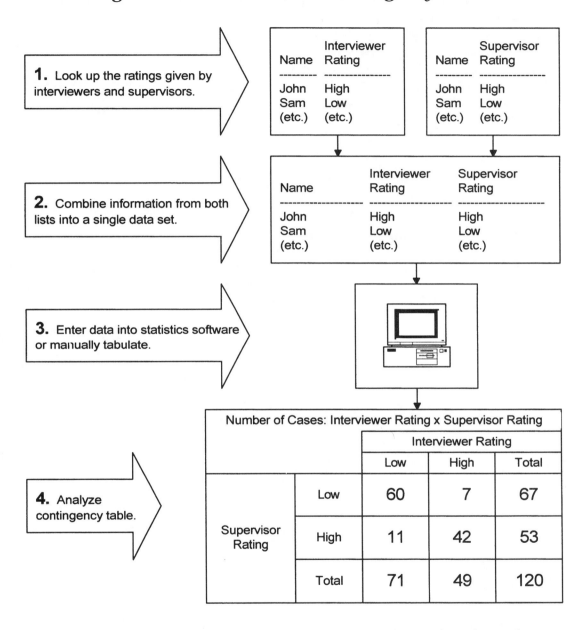

1. Look up the ratings given by interviewers and supervisors.

Name	Interviewer Rating
John	High
Sam	Low
(etc.)	(etc.)

Name	Supervisor Rating
John	High
Sam	Low
(etc.)	(etc.)

2. Combine information from both lists into a single data set.

Name	Interviewer Rating	Supervisor Rating
John	High	High
Sam	Low	Low
(etc.)	(etc.)	(etc.)

3. Enter data into statistics software or manually tabulate.

4. Analyze contingency table.

Number of Cases: Interviewer Rating x Supervisor Rating				
		Interviewer Rating		
		Low	High	Total
Supervisor Rating	Low	60	7	67
	High	11	42	53
	Total	71	49	120

Creating a simple contingency table is easy. Simply count the number of cases for each combination of values.

To create a contingency table, simply pass through the data and count the number of cases that belong in each combination of row and column (cell). Try not to use variables with many different values, because the resulting matrix is large, hard to understand, and may have an expected number of cases in some cells that is too low for the analysis to be valid (see discussion later). If necessary, collapse categories, increase the number of cases, or both. For example, consider this original classification of geographical areas: northwest, 25 cases; northeast, 1 case; southwest, 30 cases; southeast, 2 cases. The analysis would work better if categories were collapsed, like this: north, 26 cases; south, 32 cases. The original four-way classification would have produced cells with too few cases for contingency tables to work.

Exercise #22

1. Sketch out two contingency tables you could use to analyze the geographical area and product type examples on page 68.

2. From work or school, identify three research problems that could be analyzed using contingency tables. For each problem, specify two (or more) nominal variables and two (or more) possible values of each.

   a. _____

   _____

   _____

   _____

   b. _____

   _____

   _____

   _____

   c. _____

   _____

   _____

   _____

HOW TO INTERPRET CONTINGENCY TABLES

Study Avery's contingency table in Figure 9 again. It shows a strong *positive* relationship between interviewer and supervisor ratings. It would allow him to make four conclusion statements with confidence:

1. Almost all individuals rated high by the interviewer were also rated high by the supervisor.

2. Almost all individuals rated low by the interviewer were also rated low by the supervisor.

3. Very few individuals rated low by the interviewer were rated high by the supervisor.

4. Very few individuals rated high by the interviewer were rated low by the supervisor.

Exercise #23

On the contingency table in Figure 9, write the number of each conclusion (above) in the cell that supports it.

The contingency tables in Figure 10 show other possible outcomes: a positive relationship, a negative relationship, and no relationship. Study these carefully and be sure you can recognize the basic patterns. *The key consideration is whether each row and column reflects the same proportions as the total row and column.*

Figure 10: Possible Contingency Table Results

POSITIVE RELATIONSHIP Knowing either variable allows prediction of the other.	Number of Cases:		Interviewer Rating		
			Low	High	Total
	Supervisor Rating	Low	**60**	7	67
		High	11	**42**	53
		Total	71	49	120

NEGATIVE RELATIONSHIP If the value of one variable is known, the value of the other will be the opposite.	Number of Cases:		Interviewer Rating		
			Low	High	Total
	Supervisor Rating	Low	11	**42**	53
		High	**60**	7	67
		Total	71	49	120

NO RELATIONSHIP The number of cases in each cell of the table is exactly as expected.	Number of Cases:		Interviewer Rating		
			Low	High	Total
	Supervisor Rating	Low	30	30	60
		High	30	30	60
		Total	60	60	120

NO RELATIONSHIP Knowing one variable does not help predict the other.	Number of Cases:		Interviewer Rating		
			Low	High	Total
	Supervisor Rating	Low	50	50	100
		High	10	10	20
		Total	60	60	120

You don't have to be a statistician to understand contingency tables. *The key is recognizing when rows and columns do not show the same proportions as total rows and columns.*

UNDERSTANDING SIGNIFICANT RESULTS

How confident can you be that any relationship a contingency table shows is not just sample variation? After all, even if there were no actual relationship between supervisor and interviewer evaluation, *some* of the time Avery would find one anyway, just because of the particular cases he happened to select.

There are five easy steps to a solution. (You might have to read this section a couple of times. Don't worry, it is short, and an example follows.)

1. **Create a contingency table for the data.** You already know how to do this from the previous section.

2. **Create a second table, containing the expected number of cases if there were NO relationship between the two variables.** This is easy. The row, column and table totals will be the same. Calculate values for the other cells with this formula:

$$expected~\#~of~cases = \frac{row~total * column~total}{table~total}$$

The expected number of cases is important. If the expected number of cases (step 2) for any cell is below 5, chi-square analysis of a contingency table is not appropriate. Also, if any cell in a 2×2 contingency table has an expected number of cases below 10, you should consult a statistics text for corrections to the chi-square formula. It is best to avoid these problems by reclassifying data into fewer categories, or increasing the number of cases, or both.

3. **Calculate the value of a summarizing number called** *chi-square.* Chi-square (pronounced "kye") serves as a measure of the difference between the number of cases you *actually* observed and the number of cases you *expected* to observe (table #1 versus table #2). If it is low, then the overall difference was small and it is unlikely a relationship exists. If it is high, then the overall difference is large and there probably is a relatonship.

Here is the formula. It says: For each cell, subtract the *actual* number of cases from the *expected* number, square the result, and divide by the *expected* number (that is, the value in a cell in table #2 minus the value of the same cell in table #1, squared, divided by the value in table #2). Finally, total this figure for all of the cells. (Don't use row, column or table totals.)

$$x^2 = \sum \left(\frac{(actual - expected)^2}{expected} \right)$$

UNDERSTANDING SIGNIFICANT RESULTS (continued)

Can you see how this measures the strength of a relationship? If a table were greatly different (from one with no relationship between variables), chi-square would be large, wouldn't it? The larger the chi-square statistic, the more likely there is to be a strong relationship between the variables. (Notice it doesn't tell you what *kind* of relationship, just that one is out there. To understand the direction—positive, negative, none—you must scan the table.)

4. **Calculate the degrees of freedom.** The *degrees of freedom* is closely related to the number of rows and columns in the table:

degrees of freedom = (number of columns – 1) * (number of rows – 1)

(Don't count the total column or the total row.) All of the tables in this book are 2×2, so for these tables, the *degrees of freedom* is always (2–1)*(2–1) or 1. We use this value in the next step.

5. **Look up the probabilty.** Think about a large population where there is no relationship at all between the two variables you are measuring. If you were to select many samples from this population, create contingency tables and calculate chi-square for each, you could create a frequency distribution. It would show how often you would expect to find a relationship (when there wasn't one), just because of the cases you happened to select. That probability would represent the odds that chi-square would be this big (or bigger) if, in reality, there were no relationship. If it were 5% or below, the results would be *statistically significant*. Statistically significant means it is unlikely the particular cases you happen to pick cause the difference. In other words, the usual variation between samples would cause a difference this big less than 5% of the time, if there truly were no difference.

These values are published in most statistics books as tables of the chi-square distribution. For our purposes here, let's simply note that for one degree of freedom, chi-square needs to be larger than 3.8 to have a probability below 5%. That is, for a 2×2 contingency table, a chi-square larger than 3.8 (with no relationship between the variables) would occur less than 5% of the time.

Here is a simplified table of values of the chi-square distribution. All the examples in this book use the 0.05 level of significance and a 2 × 2 table, so the appropriate chi-square value for them is always approximately 3.8. Level of significance is just another way of saying "statistically significant." As we just discussed, it means the usual variation between samples would produce a chi-square this large (or larger) only five times out of 100, if there really were no relationship between the variables. Still another way to say the same thing is to say you are 95% confident that the contingency table indicates an association between the variables.

Rows	Columns	Degrees of Freedom (Rows-1) times (Columns-1)	Critical Value of Chi-square for the .05 Level of Significance	Critical Value of Chi-square for the .01 Level of Significance
2	2	1	3.8	6.6
3	2	2	6.0	9.2
3	3	4	9.5	13.3

(Remember, the values in this table have been rounded for simplicity. If your project requires more precise values, consult the more detailed tables in any college statistics text.)

The term *degrees of freedom* is used with many different statistical techniques and the formula is different for each. Practically, it usually serves to point the researcher to the proper probability distribution to use. (In this case, it points to the proper chi-square distributions—there is a different one for each different *degree of freedom*.)

Although statistical software is an easier approach, it is not difficult to analyze contingency tables by hand. Figure 11 might give you an idea for a spreadsheet template. Following the example is an exercise to cement your understanding of this process.

Figure 11: Significance Calculations

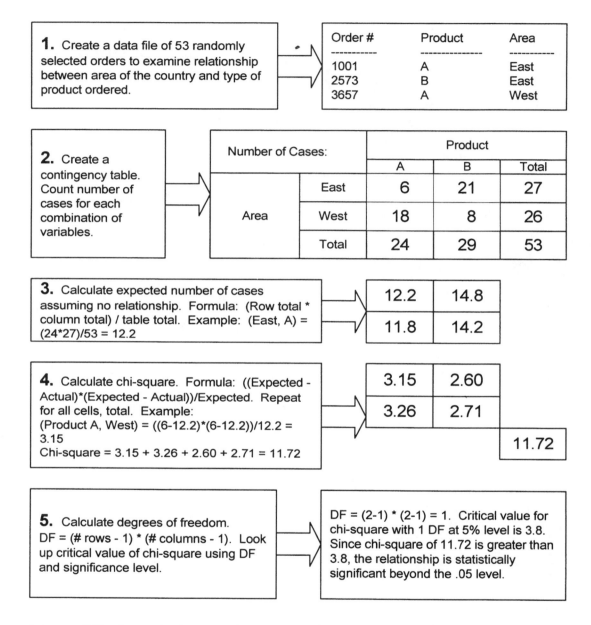

1. Create a data file of 53 randomly selected orders to examine relationship between area of the country and type of product ordered.

Order #	Product	Area
1001	A	East
2573	B	East
3657	A	West

2. Create a contingency table. Count number of cases for each combination of variables.

Number of Cases:		Product		
		A	B	Total
Area	East	6	21	27
	West	18	8	26
	Total	24	29	53

3. Calculate expected number of cases assuming no relationship. Formula: (Row total * column total) / table total. Example: (East, A) = (24*27)/53 = 12.2

12.2	14.8
11.8	14.2

4. Calculate chi-square. Formula: ((Expected - Actual)*(Expected - Actual))/Expected. Repeat for all cells, total. Example: (Product A, West) = ((6-12.2)*(6-12.2))/12.2 = 3.15
Chi-square = 3.15 + 3.26 + 2.60 + 2.71 = 11.72

3.15	2.60
3.26	2.71

11.72

5. Calculate degrees of freedom. DF = (# rows - 1) * (# columns - 1). Look up critical value of chi-square using DF and significance level.

DF = (2-1) * (2-1) = 1. Critical value for chi-square with 1 DF at 5% level is 3.8. Since chi-square of 11.72 is greater than 3.8, the relationship is statistically significant beyond the .05 level.

It is not difficult to calculate significant results for a contingency table. You might consider creating a spreadsheet template to handle simple contingency tables. (Values in step #3 and step #4 have been rounded. If you duplicate these problems, your values might be slightly different due to differences in rounding.)

Figure 12: Exercise #24

Try your skill on this problem. Write your answers on this figure. Refer to the previous figure as necessary.

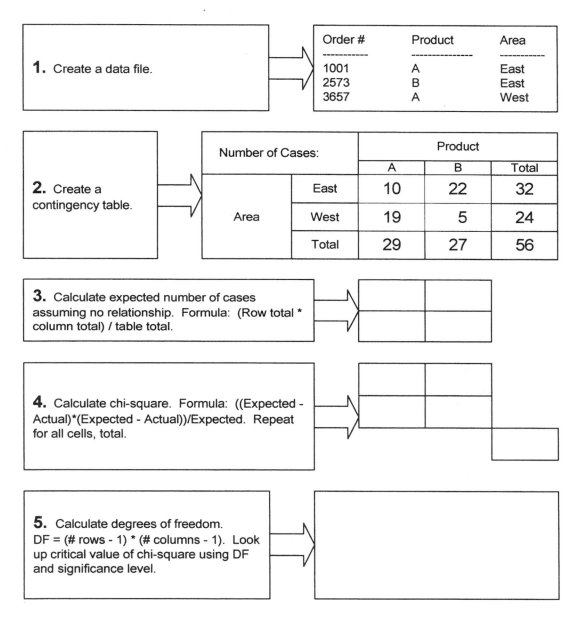

1. Create a data file.

Order #	Product	Area
1001	A	East
2573	B	East
3657	A	West

2. Create a contingency table.

Number of Cases:		Product		
		A	B	Total
Area	East	10	22	32
	West	19	5	24
	Total	29	27	56

3. Calculate expected number of cases assuming no relationship. Formula: (Row total * column total) / table total.

4. Calculate chi-square. Formula: ((Expected - Actual)*(Expected - Actual))/Expected. Repeat for all cells, total.

5. Calculate degrees of freedom.
DF = (# rows - 1) * (# columns - 1). Look up critical value of chi-square using DF and significance level.

(The answers are on page 114.)

Summary

Contingency tables are statistical tools you can use to discover relationships between two or more variables containing nominal data. You should be able to recognize tables that show positive, negative, and no relationship. Chi-square is a summarizing number that can be calculated from a contingency table. The chi-square distribution tells whether a relationship suggested by a contingency table is best explained by expected sample variation or by a genuine relationship between variables.

. . . Avery constructed a contingency table of supervisor and interviewer ratings of promotability to summarize the study of 120 randomly selected new hires. The sample revealed a strong positive relationship between interviewer ratings and supervisor ratings, statistically significant at the .05 level.

. . . "Well," Avery thought, "at least I am 95% confident there is some positive association between interviewer and supervisor ratings. It appears the interviewers really know what the supervisors want. Maybe we'd better take a look at ways to increase the number of promotable individuals our interviewers see at the start of this pipeline." . . .

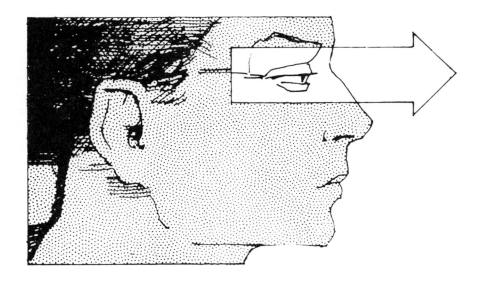

SECTION 9
Linear Regression

Hiram, a small-business owner, was planning several years in the future. "It looks like we have some interesting times ahead of us," he thought. Because of outside changes, production was going to fluctuate wildly. "What's going to happen to our costs?" he wondered. He turned to his accountant, "I know information is still tough to get with the new computer system, but we know total costs and our production in units for every month for the last three years. Our expense structure has been fairly stable during that time. What can we do to estimate future costs?"

What would you do?

LINEAR REGRESSION

Requirements: Two interval or ratio variables

Purpose: Uncover linear relationships between concepts. Given value(s) of one variable, estimate value(s) of the other. Estimate strength and direction of relationship with correlation coefficient.

WHAT IS LINEAR REGRESSION?

Linear regression is a tool that can uncover linear relationships between two interval or ratio variables. The goal is to condense the information in both variables into the mathematical model of a straight line. With such a formula, if you know one variable, you can predict the other. The model for a straight line is

$$y = (m * x) + b$$

where

- y = *the value of one variable* (the dependent one)

- x = *the value of the other variable* (the independent one)

- m = *the slope of the line* (If it's positive, the line slopes upward from left to right. If it's negative, the line slopes downward from left to right. The farther from zero, the more nearly vertical the line. The closer to zero, the more nearly horizontal. If this is unclear, you might want to skip ahead to Figure 14. The Positive Relationship chart shows a line with an m that is positive. The Negative Relationship chart shows a line with an m that is negative. The No Relationship chart shows a line with an m that is close to zero.)

- b = *a constant* (It's also the place on the y-axis where the line crosses.)

Hiram's problem can be modeled by a line. The dependent variable (y-axis) is total cost. The independent variable (x-axis) is the number of units sold. The slope is the variable cost of producing one unit. The constant represents the cost when no units are sold—total overhead. Expressed as a linear equation:

*total cost = (cost per unit * number of units) + total overhead cost*

To solve this problem, Hiram should create a data set where each month is a case and each case contains the variables *total cost* and *number of units*. Feeding that data through a regression analysis will produce an equation. The line this equation produces is called a *regression line*. It represents the best fit of all the points around a single line. You can see examples of regression lines in Figures 13 and 14.

Neither Hiram nor anyone else does linear regression problems by hand (although it is possible for simple ones). Statistical and spreadsheet software contain very easy regression procedures. Simply start the regression procedure. Then, identify the data for each variable (either by variable name for statistical software, or by specifying the range for spreadsheets). With that information, most software will produce scatter graphs (to be discussed shortly) and calculate the values of m, x, and b.

Exercise #25

To cement your knowledge of the basic model of a line, calculate y given the information below. Also state whether the relationship between variables is positive or negative. (Positive means the line slopes upward—more of one variable means more of the other. Negative means the line slopes downward—more of one variable means less of the other.)

m	x	b	y $y = (m*x)+b$	Positive or Negative Relationship?
–2	3	10	4	Negative
3	2	15		
5	5	–10		
–4	6	3		
–1	1	5		
4	2	8		

(*The answers are on page 115.*)

You may have recognized that linear regression analysis is similar to contingency table analysis (Section 8). Both try to uncover relationships between two or more variables. However, contingency tables use nominal data; regression analysis uses interval or ratio data (Section 3). Do you see how the additional information in interval or ratio data provides a much more complete picture of the relationship between variables?

SCATTER GRAPHS

To understand linear regression, you should also understand a special kind of graph called a *scatter graph*. They have the following characteristics:

- The vertical axis (y-axis) represents one variable.

- The horizontal axis (x-axis) represents the other.

- Each case is represented by a dot, a letter or another symbol. The vertical distance is determined by that case's score on the *y* variable. The horizontal distance is determined by that case's score on the *x* variable.

Figure 13 shows the solution to Hiram's problem and its associated scatter graph.

Although only three cases are shown from the data file, each one of them corresponds to a point on the graph. The other points represent the rest of the cases. As you can see, to forecast future costs, all he would have to do is plug in the *planned unit volume* for a given month and solve the equation for *total cost*. For example, if he knew that production in a future month would be 1,000 units, he could calculate *total cost* for that month like this:

$$m = \text{unit cost} = \$67.36$$

$$b = \text{total overhead cost} = \$8,608$$

$$x = \text{planned unit volume} = 1,000 \text{ units}$$

$$y = (m*x)+b$$

total cost = (unit cost * unit volume) + total overhead

total cost = ($67.36 * 1,000) + $8,608

total cost = $75,968

By providing estimates of *m* and *b*, regression analysis lets you forecast *y* for any value of *x* you choose. The accuracy of those forecasts will depend on how well a line can model the relationship, how accurate the data is and how representative past data will be of future situations.

Figure 13: Example of Regression Analysis

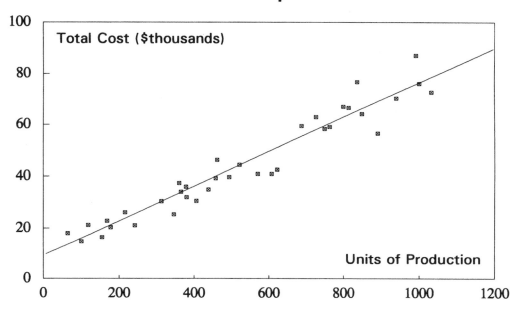

Scatter Graph

Data File

Variables		
Month	Units	Cost
January	360	37,374
February	727	63,006
March	608	40,970
(etc.)	(etc.)	(etc.)

Results

Output Variable	Value
m	67.36
b	8,608
r	0.96
r squared	0.93

Conclusion

Total Cost = ($67.36 * Unit Volume) + $8,608. The relationship explains 93% of the variation.

Exercise #26

Estimate total cost for Hiram for each of the following production levels: 1,500 units; 100 units; 250 units.

(The answers are on page 115.)

Often, just looking at a scatter graph gives an intuitive understanding of the relationship between variables. Four examples are shown in Figure 14.

- A strong *positive* relationship. Although all the points are not *exactly* on a straight line, the regression line provides a good estimate of them. As the scores on one variable increase, so do the scores on the other.

- A strong *negative* relationship. In this graph, as scores on one variable increase, scores on the other decrease.

- *No relationship*. The value of one variable is just not a good way to predict the other.

- A *nonlinear* relationship. This is a strong relationship that cannot be modeled by a line. As the scores on the *x* variable increase, the *y* scores first increase, then level off, then decrease. Linear regression would be the wrong statistical model for this relationship because a line cannot properly model the obvious relationship.

Linear regression works best in situations where scatter graphs look like the first two examples. It's a good idea to examine a scatter graph of your data to see if linear regression is an appropriate model. Also review the correlation coefficient (see next section) as an additional check on the overall accuracy of the predictions.

Figure 14: Examples of Possible Relationships

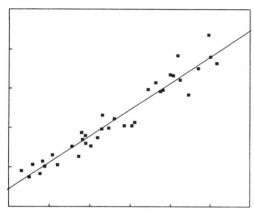

Positive Relationship
r = .97

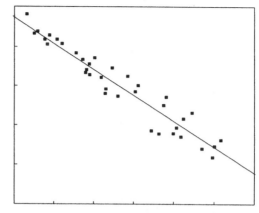

Negative Relationship
r = .96

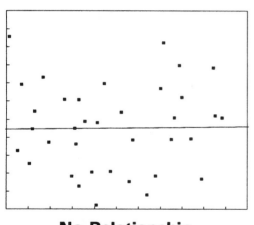

No Relationship
r = .02

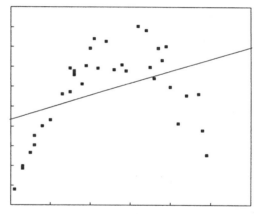

Nonlinear Relationship
r = .40

CORRELATION

Most software produces a useful summarizing number associated with linear regression, the *correlation coefficient*. Its symbol is the letter r.

The correlation coefficient shows the *strength* of the association between the variables. It ranges between +1 and –1.

- A value of +1 suggests a perfect *positive* association between the concepts. As one increases, so does the other, in perfect proportion. There is no difference between what the line predicts and what the real world produces.

- A value of 0 suggests *no association* between the concepts. If you know the value of one concept, it has no predictive value in calculating the value of the other.

- A value of –1 indicates a perfect *negative* association between the variables. As one increases, the other decreases, in perfect proportion. Again, there is no difference between what the line estimates and what the real world produces.

The correlation coefficient is noted on the four scatter graphs in Figure 14. Can you see how the correlation coefficient measures how tightly the points swarm around each regression line?

The *square* of the correlation coefficient estimates how much of the total variation is explained by the relationship. For example, if *discretionary spending* and *total income* have a correlation coefficient of 0.9, *total income* explains 81% of total variation in *discretionary spending* (0.9 * 0.9 = 81%); 9%, (100% – 81%), would be caused by other (unknown) factors.

Exercise #27

How much of the total variance is explained by the relationship, given the following correlation coefficients?

Correlation Coefficient	%Total Variance Explained	Positive or Negative Relationship
.9	81%	Positive
−.5		
−.25		
.75		
−.3		
.6		

(The answers are on page 115.)

Summary

Linear regression is a statistical process that uses the standard formula for a line, $y = (m*x)+b$, as a model for a relationship between two concepts. A scatter graph shows how individual cases score on two variables. The regression line is the line that minimizes the square of the differences between actual and predicted values. The correlation coefficient is a summarizing number that ranges from +1 to −1 and shows the strength and direction of the relationship. Values close to +1 suggest a strong positive relationship. Values close to −1 suggest a strong negative relationship, and values close to zero suggest no relationship. The square of the correlation coefficient indicates the percent of total variation explained by the relationship between the two variables.

> . . . Hiram created a regression analysis of total cost and unit volume. "From this analysis," he thought, "I see we have a fixed overhead of around $8,600 per month and a variable cost of around $67 per unit. Since I have a list of projected unit sales by month, this relationship will let me forecast total cost by month, because I don't anticipate any major changes from this history. I am pretty confident of the results, because historically, these estimates of volume and overhead account for over 90% of the total variation of costs." . . .

SECTION 10

Confidence Intervals

It was time for the monthly review of operations. The hot button this month was customer satisfaction. Sam, the Operations Manager, expected Naomi, his boss, to pose some challenging questions. Sam felt confident, though. Randomly selected customers completed a survey each month. Their answers were combined into a "satisfaction index" (100 = perfectly satisfied, 0 = totally dissatisfied). Sam began the meeting with, "Our customer satisfaction index averaged 85 for last month, which is really quite good. . ."

Naomi interrupted, "Quite good? You only surveyed 50 out of 5,000 customers. I'll ask another 50 and I bet I get something completely different. Probably worse!"

If you were Sam, what would you do?

CONFIDENCE INTERVALS

Requirements:

- Sample measuring a ratio or interval variable
- Frequency distribution of variable does not have to be normal if number of cases > 30

Purpose: Specify a range in which the population mean probably lies.

Distribution: Standard Normal for cases > 30, otherwise t-distribution (if frequency distribution is roughly bell-shaped).

WHAT ARE CONFIDENCE INTERVALS?

This chapter shows how to estimate the mean of a population by studying a sample. From Section 4, you will remember that researchers study samples because studying the entire population is usually not feasible. However, samples create a problem. Of the many different sample means, which is the best estimate of the true (population) mean?

The solution is a *confidence interval*. A confidence interval is a statement such as "I am 95% confident that the true (or population) customer satisfaction rating is between 75 and 80." It consists of two parts: a *range* of possible values (75–80) and a *probability figure* (95%). The probability figure states the odds that the range contains the actual population value.

Exercise #28

Identify an example of a confidence interval that would answer a useful question at work or school. What level of confidence would you select? How broad could the range be and still be useful?

_____

_____

Characteristics of Sample Means

Imagine walking down a busy street. Let's say that you measure the height of the first 10 people you meet. Then, you measure the next 10, and the next, until you have measured 10 groups. Would you expect the mean height for each group to be the same?

Of course not. It would vary, *but in predictable ways*. You would expect *most* of the sample means to be *fairly close together*. You would expect sample means to vary *less* than individual measurements (because in averaging, extreme individual measurements cancel one another out). You would also expect the average height of all 100 people to be a *more accurate* indicator of the population average than any single average from any of the groups of 10.

These insights are true for all groups. Specifically:

- Group averages vary less than individual measurements.

- The expected value of a sample mean is the population mean.

- The more cases studied in a sample, the more precise the estimate.

Figure 15: Two Related Distributions

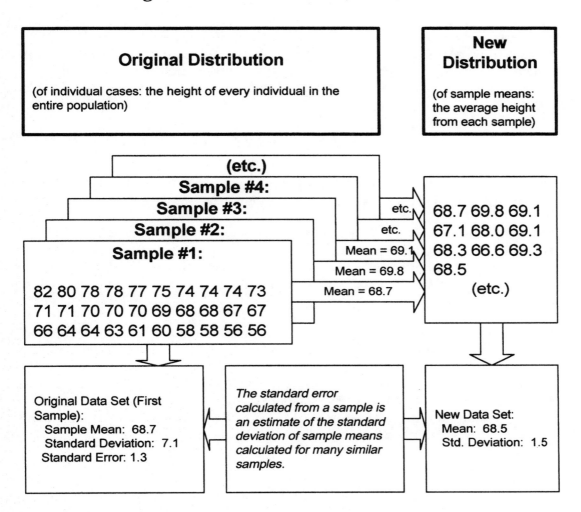

A single sample of adequate size and quality contains enough information to estimate what the distribution of sample means looks like. Compare the standard error of the first sample (1.3) to the standard deviation of sample means (1.5). The estimate is close, isn't it?

THE DISTRIBUTION OF SAMPLE MEANS

In the last example, a case was an individual (whose height we were measuring). In that study, we selected ten samples. So, we could create a *new* data set (with ten cases), where *each sample mean is a case.* The distribution of these sample means across the range of their possible values would be a *new* distribution—the *distribution of sample means.*

Exercise #29

Study the previous figure carefully. Are you clear about the difference between the *original* distribution of people and the *new* distribution of sample means?

Most of the time, you will not run a study on many samples—usually there is just one. However, even one good random sample contains enough information to estimate what this new distribution looks like. When there are over 30 cases in your data set, you can model the distribution of sample means using a normal distribution with the following characteristics:

► **MEAN:** The same as the sample mean. The mean of the sample is the best estimate you have, and over time the average of all such sample means is the population mean.

► **STANDARD DEVIATION:** The standard deviation of this new distribution is the standard deviation of the sample divided by the square root of the number of cases. This is also called the *standard error* (of the mean).

$$s_{\overline{x}} = \frac{s}{\sqrt{n}}$$

If there are 30 or more cases, it really doesn't matter if the original distribution is a normal distribution or not. (For example, instead of a sand pile, the frequency distribution of the variable might look like several sand piles poured against one another, or any other of an infinite number of different shapes.) The important idea is that regardless of the shape of the original frequency distribution, when the number of cases in a random sample is greater than 30, the *new* distribution, the distribution of sample means, can still be modeled by the normal distribution.

HOW TO CREATE A CONFIDENCE INTERVAL

The normal curve can serve as a model for this distribution of sample means. (Don't become confused—the normal curve might also be a good model for the original distribution of individual cases, but that is *not* the distribution we are using.) It can provide the probability that the population mean falls within a particular range. You can create such a confidence interval in three easy steps.

STEP 1: **Calculate the mean and standard deviation for the sample.** (Review Section 5, if necessary.) For this example, let's say a random sample of 50 customers produced a mean satisfaction index of 85 with a standard deviation of 5. (Be sure to use the formula for a *sample* standard deviation.)

STEP 2: **Calculate the standard error of the mean.** Use the formula on the previous page:

$$s_{\bar{x}} = \frac{s}{\sqrt{n}}$$

$$s_{\bar{x}} = \frac{5}{\sqrt{50}}$$

$$s_{\bar{x}} = 0.71$$

Compare the standard deviation of individual cases (5) to the standard deviation of the sample mean (0.71). Do you see how the averaging process reduces the variability of the sample mean?

STEP 3: **Calculate critical values.** This is exactly the same process you used in Section 7. Let's say you want to be 95% confident that your confidence interval contains the actual population mean. The critical points are plus and minus 1.96 standard deviations (per standard normal table in Section 7). Using the same formulas we used earlier (Section 7):

$$x = (z * \text{std. error}) + \bar{x} \qquad x = (z * \text{std. error}) + \bar{x}$$

$$x = (-1.96 * 0.71) + 85 \qquad x = (1.96 * 0.71) + 85$$

$$x = 83.6 \qquad\qquad x = 86.4$$

(You will notice, however, that instead of *s*, the sample standard deviation, we are using the *standard error*, the standard deviation of the new distribution of sample means.) So, for this example we can be 95% confident that the true population mean is somewhere between 83.6 and 86.4.

Exercise #30

Test your understanding with these questions:

1. Given the following information, calculate the standard error. The first line is the same as the example you just studied. (Critical points for a two-tailed test are plus and minus 1.96 for 95%, and plus and minus 2.58 for 99%.)

Mean $\bar{x}$	Std. Dev. s	# Cases n	Standard Error $s_{\bar{x}} = \dfrac{s}{\sqrt{(n)}}$	Confidence Interval @ 95% $(z * s_{\bar{x}}) + \bar{x}$ to $(-z * s_{\bar{x}}) + \bar{x}$	Confidence Interval @ 99% $(z * s_{\bar{x}}) + \bar{x}$ to $(-z * s_{\bar{x}}) + \bar{x}$
85	5	50	0.71	83.6 to 86.4	83.2 to 86.8
85	3	35			
100	10	30			
500	50	65			
20	2	40			
50	5	37			

2. State the confidence intervals as a sentence. Example: "I am 95% confident that the population mean is between 83.6 and 86.4." (Or, "This sample indicates that there is a 95% chance that the population mean falls between 83.6 and 86.4.")

_____

_____

_____

3. A random sample of 65 production lots had a mean percent defective of 5% with a standard deviation of .5%. Express your estimate of the true percent defective as a confidence interval at 99%. _____

_____

_____

_____

4. On a scale of 1 to 10, a random sample of 50 employees (from a large manufacturing plant) rated job satisfaction at an average of 6.5 with a standard deviation of 1. Express your estimate of the true value for all employees at the plant at the 95% confidence level. _____

_____

_____

_____

5. At your restaurant, a random sample of 80 guest checks provided a mean total sales of $25 with a standard deviation of $2. Express your best estimate of the actual average guest check at the 95% confidence level. _____

_____

_____

_____

(The answers to these questions are on page 115.)

THE t-DISTRIBUTION

Until now, everything you have learned in this section applies to samples of 30 or more. When your sample is smaller, you must make one easy change. You must substitute a t-distribution for the z-(standard normal) distribution. Also, to use the t-distribution this way, you must be able to assume that the original distribution is roughly bell-shaped.

Let's review the example in the opening story, which measured a satisfaction index of 85. Earlier we decided that, on the basis of a sample of 50, Sam could be 95% confident that the true satisfaction index was between 83.6 and 86.4. What if there were only 6 cases in the sample instead of 50? We would do the same calculation, except that the standard error would be larger and instead of using values from the z-distribution, we would use the appropriate t-distribution. In this case, we would use the one for 5 degrees of freedom, (df = $n-1$ = 6–1 = 5). Here is a simplified table of values from selected t-distributions. These values are appropriate for confidence intervals because they are two-tailed (Section 7, page 62).

Degrees of Freedom ($n-1$)	Value of t for 0.05 Level of Significance	Value of t for 0.01 Level of Significance
5	2.57	4.03
10	2.23	3.17
15	2.13	2.95
20	2.09	2.85
25	2.06	2.79

This table says that for a t-distribution with 5 degress of freedom, 95% of the area is between –2.57 and +2.57 standard deviations from the mean. Ninety-nine percent of the area is between –4.03 and +4.03 standard deviations.

The only difference between the earlier example and this solution is that the standard error is larger and we plug in values from the appropriate t-distribution.

$$s_{\bar{x}} = \frac{s}{\sqrt{n}} \qquad x = (t * \text{std. error}) + \bar{x} \qquad x = (t * \text{std. error}) + \bar{x}$$

$$s_{\bar{x}} = \frac{5}{\sqrt{6}} \qquad x = (-2.57 * 2.04) + 85 \qquad x = (2.57 * 2.04) + 85$$

$$s_{\bar{x}} = 2.04 \qquad x = 79.7 \qquad x = 90.2$$

Compare this to the earlier example on page 93. Do you see how the t-distribution increases the size of the confidence interval?

The t-distributions are very similar to the z-distribution. They, too, count standard deviations from the mean. However, there is a different t-distribution for every possible value of *degrees of freedom*. (In this test, *degrees of freedom* [df] equals the number of cases less one.) Above 30 df, all the different t-distributions are very close to the z-distribution. Below 30 df, each successive t-distribution looks like a normal distribution that has been increasingly "stretched." More and more of the middle of the distribution flows into the tails. Practically, this means studies with fewer cases require larger intervals to produce the same level of confidence. After all, fewer cases mean less information, so statistics *should* be less precise. Think of the t-distribution as a series of distributions, similar to the normal distribution, but used with samples of less than 30 cases.

Summary

Confidence intervals allow you to estimate population means from sample means. They consist of a range of possible values and a probability figure that represents the likelihood that the range contains the actual population value. The broader the range, the higher the probability that it contains the actual population value. The greater the number of cases in the sample, the narrower the range. The probability associated with confidence intervals can be derived from the normal distribution when the number of cases is 30 or above. For samples with fewer than 30 cases, but where the original distribution can be assumed to be bell-shaped, the calculation is the same, but substitute one of the t-distributions for the z-distribution.

> . . . Sam continued, "Let me explain. Based on the variability in the sample, I am 95% confident that the true customer satisfaction index is between 80 and 90, which is a very respectable rating." . . .

SECTION 11

Differences in Sample Means

The warehouse David supervised was testing a new dispatch system for delivery trucks. The company wanted to use it nationwide, but only if the new system would allow at least 10 more stops per truck. Otherwise, it wasn't worth the investment. Now, the V.P. of Distribution had just told him the decision would be made tomorrow and both their careers were on the line. The trip reports listed the number of stops and which dispatch system was used. The dispatch clerk could assemble a random sample of trip reports from each system.

"Now what?" thought David.

What would you do?

DIFFERENCES IN SAMPLE MEANS

Requirements:

- One random sample from each of two different large populations, measuring a ratio or interval variable
- Samples are unrelated (independent)
- Number of cases in each sample > 30

Purpose: Estimate the probability that the population means are different by some amount.

Distribution: Standard Normal

ABOUT DIFFERENCES IN SAMPLE MEANS

David needs to analyze a difference in two population means by using samples. The two populations are the set of all trips under the old system and the set of all trips under the new system. If he could be confident that the actual average number of trips under the new system were at least 10 more than under the old, the decision would be clear.

This technique solves many common problems. For example:

- A manufacturer installs two machines in a production process. Each takes a different amount of time to complete a job, but on the average, the vendor claims one will be five minutes faster per production lot. The company wants to compare average time per job using machine A to average time per job using machine B.

- A retail store has uncovered a way to separate customers into two groups. It wants to compare average advertising recognition scores from group A to group B.

- A company tests a new product in two different cities, to see if there is a significant difference between sales in the two different locations.

Such questions require two variables in the data set. One variable measures interval or ratio data (production time, advertising recognition scores, sales, number of delivery stops, etc.). A second, nominal variable, splits the cases into two groups (machine A or B, group A or B, city A or B, old dispatch or new dispatch, etc.).

Exercise #31

On a separate sheet of paper, list three tests for significant differences in sample means that you could use at work or school. For each, state the interval or ratio variable you are measuring and the nominal variable that defines the two groups.

A SIMPLE QUESTION AND A COMPLICATED DECISION

Statistically, questions like these can be answered by a *test for significant differences between sample means*. Unfortunately, though the kind of questions that are appropriate for this technique are obvious, the right statistical tool to use is not. There are several different tools. Selecting the appropriate one depends on meeting the conditions associated with each.

The technique that follows (a *z-test for a significant difference in sample means*) is appropriate when the following conditions are met:

1. The two random samples come from two different, large populations.

2. The samples are independent. (There is no relationship between cases in one sample and cases in the other.)

3. Each sample contains more than 30 cases.

You can also use this technique with samples of less than 30 if the interval/ratio variable can be assumed to have a normal frequency distribution and you happen to know the actual variance of the populations. Just substitute the actual variances for the sample variances in the formula on page 102.

WHEN DOES A DIFFERENCE COUNT?

Questions about differences in means can be difficult because there will usually be *some* difference between the means of *any* two samples. How big must a difference between samples be, to suggest an actual difference between the populations?

The solution is the *probability* that a difference of a certain size is due to random variation between samples. If that probability is low, then it is likely the population means are truly different. If the three conditions listed earlier are satisfied, you can read that probability from the standard normal curve, using the formula from the example that follows.

Example. Using a spreadsheet, David quickly computed the mean and variance for the number of stops made on each trip, using two random samples of trip reports (one from each dispatch system). Here are the results. Although he found an average difference above 10 orders per load, David wanted to be 95% confident that it was not due to the trip reports he happened to select.

Dispatch System	Mean	Variance	Number of Cases
New	57	12	45
Old	45	10	50

The Recipe. To solve this problem, simply plug the appropriate numbers in the formula on the right and calculate the value of z. A standard normal table, as shown in Figure 7, says that only 1% of the distribution is in the area greater than 2.33. Since 2.93 is greater than 2.33, this difference is significant beyond the .01 level. Less than one time out of 100 would a difference this great or greater occur, if the difference were truly less than 10. David can be 99% confident there is at least a 10 stop difference between the old and new dispatch system.

$$z = \frac{(\bar{x}_1 - \bar{x}_2) - difference}{\sqrt{\dfrac{(s_1^2)}{n_1} + \dfrac{(s_2^2)}{n_2}}}$$

$$z = \frac{(57 - 45) - 10}{\sqrt{\dfrac{12}{45} + \dfrac{10}{50}}}$$

$$z = 2.93$$

When using the formula, be careful when you substitute values into the upper part of the fraction. Although not obvious from the formula, it is easy to switch the means around or to accidentally change the sign of the "difference." David's problem is typical of the kind you are likely to meet: He expected the new system to be better than the old, he had an idea of how much better it had to be, and his sample results confirmed that expectation. He just wanted to know how confident he could be that the results were strong enough. If the samples had shown the new system to be *worse*, he wouldn't have even bothered with the calculations—the odds the new system was actually better than the old, even though the samples said it was not, would be below 50%—but that was not the case. So, he subtracted the old from the new (57 minus 45). Then he reduced that figure by the minimum he could accept (by subtracting 10). For your problems, you should do something similar, and follow the example carefully. If you switch the means around, or change the sign of the difference, the process still has intuitive meaning, but it can be mind-twisting to figure out. Follow the example faithfully—and be careful.

Exercise #32

Assume David found different results for the test. Test your skill by calculating the value of z for the following results. The critical value for the z-score @ 5%) is 1.65.

Dispatch System	Mean	Variance	Number of Cases	Difference	z	Significant @ .05 Level?
New	55	20	40	5		
Old	50	10	40			
New	20	1	31	1		
Old	18	2	31			
New	7	3.5	50	0		
Old	6	3.5	50			

(The answers to this question are on page 115.)

WHAT ARE YOU REALLY DOING?

Although the recipe looks simple, it is important to understand what it does. Imagine selecting repeated samples, like the ones you used, from two very large populations whose means actually differed by the amount you are testing. If, for each pair of samples, you calculated the difference of sample means, you could create a frequency distribution. It would show the relative frequency of an *apparent* difference of two *sample* means when the *actual* difference in the *population* means was the figure you chose. That distribution would tell you how often differences of a certain size could be expected to occur, given the real difference in the populations. You are using the standard normal distribution as a model for this distribution of the difference of two sample means.

OTHER TECHNIQUES

If your problem doesn't meet the conditions for a *z-test for significant difference in sample means*, you will need to consult a statistics text. Although we will not discuss these tools in this book, you might check these topics first, in a more advanced text.

- If there are less than 30 cases, and if you do not know the population variance figures, and if you can assume the interval/ratio variable has a normal frequency distribution, look up the *t-test for the difference of two means*. (There are two different t-tests, based on different assumptions about the variance, so be sure to use the right one.)

- If you can't assume normality, and there are fewer than 30 cases, look up *non-parametric tests*.

- If the samples are not independent, look up *paired observation tests*.

- You might also be able to categorize the interval or ratio information into a nominal variable, and use the *contingency table* techniques discussed earlier in this book.

A FINAL WORD ABOUT SIGNIFICANCE

Statistically significant results are important in the academic world, and they are always the desired goal in applied business research. However, in business, sometimes less than ideal results can still have value, and even statistically significant results must be examined for practical significance.

Scholars have the luxury of not replacing existing knowledge unless they feel very certain new information is correct. So, they always start from the same assumption, the *null hypothesis:* existing information is correct and research will not uncover anything different. To be published, a scholar must reject some form of this null hypothesis, and suggest that an *alternative* hypothesis is better. Results must be statistically significant at the .05 level or beyond. This strategy *minimizes* the probability of a type I error (alpha), the odds scholars will accept the alternative hypothesis (change existing knowledge) when they shouldn't. However, it *maximizes* probability of a type II error (beta), the odds that scholars accept the null hypothesis (fail to change existing knowledge) when they shouldn't. The type II sacrifice is acceptable, though, because existing knowledge, theoretically, has already been held to a high standard of proof.

In business, the situation is somewhat different. The null hypothesis is the action executives would take based on judgment alone; the alternative hypothesis is another course of action. Sometimes executives have enough time and other resources to act like scholars: They can demand a high standard of truth before they change their initial, pre-research judgment. However, other situations require amplifying and acting on weak, uncertain signals. When one has little confidence in his or her initial judgment, research results at any *higher* level of confidence are an improvement—not necessarily ideal, not necessarily a goal to aspire to, but possibly illuminating enough to make a difference. Unlike scholars, executives don't always have the luxury of a high level of confidence in their default position, nor do they always have the luxury of postponing action until ideal results are available. In your research, always look beyond the strong, obvious results. Be sensitive to the hints and suggestions that, when combined with additional research or confirmed by additional judgment, might give an unsuspected basis for action.

Even when results are *statistically* significant, executives should ask if results are also *practically* significant: strong enough to justify *doing* something different.

Summary

Analyzing the difference of two sample means is a very useful tool. Samples from two different groups usually have different averages, which could suggest two different things: a genuine difference in the two populations, or the usual variation between samples. If the number of cases in each sample is greater than 30, and if the samples are independent and drawn from large populations, the z-distribution can be a useful tool. It allows you to make statements such as, "I am 95% confident that there is at least a 10-unit increase in the average of group X versus that of group Y."

. . . David recommended using the new system. He was able to conclude that there was less than 1 chance in 100 that the actual advantage was less than 10 stops . . .

P A R T

IV

Putting It All Together

Section 12

Now What?

Miriam was a regional vice-president in charge of four districts. One district had developed an alarming decay in sales, and it appeared that none of the typical problems was responsible for the unfortunate slump.

What would you do?

PUTTING YOUR SKILLS TO WORK

This book has covered a lot of ground. For a quick review, let's see how Miriam might put all of this together. (This does not represent an actual study. In a similar situation, your firm might need to take different steps.)

First, Miriam considered the six hurdles and decided to conduct research (Section 1). Then, she developed a theory that sales was a function of the professionalism of employees, attractiveness of the stores, price, and location. She operationally defined these concepts through reported sales and the results of a customer survey (Section 2).

On the survey, Miriam asked several questions for each concept she wanted to measure. She designed a data file to record the answers and created a variable for each question. She also added several other variables to help analyze the information: district, store number, gender of customers, zip code, etc. (Section 3). Then she tested the survey and refined it to reduce possible bias. She also selected a random sample of customers and ran similar surveys in other districts as a control (Section 4).

When the surveys were complete, she calculated the mean and standard deviation for each question and noted them on a blank survey form. As she studied it, she thought, "Now I'm looking at the typical customer who comes through my company's door" (Section 5). She created a frequency distribution to visualize how far customers lived from the store (Section 6). She used frequency analysis to estimate the actual number of dissatisfied customers in the trade area (Section 7).

In the unsuccessful district, contingency tables revealed a significant difference between old and new customers and willingness to shop at the store (Section 8). Linear regression uncovered a relationship between household income and frequency of purchases. The high slope of the regression line suggested customers perceived the product to be more of a luxury than a necessity (Section 9). She also created confidence intervals to estimate the actual population values for appropriate questions (Section 10). Finally, she discovered a significant difference in average perceived attractiveness of retail outlets among successful and unsuccessful districts (Section 11).

At the next management meeting, Miriam made several significant decisions. She increased the capital budget to improve the appearance of stores in the faltering district. She revised the advertising strategy to position the product less as a luxury and more as a necessity. She also modified the direct-mail campaign to respond to the different needs of old and new customers. Finally, she decided to relocate one store to improve access by the customers.

. . . "Well," thought Miriam. "I hadn't thought of half of this. It's been a lot of work, but I think for the first time we really know what's going on out there." . . .

Exercise #33

1. Select a research problem from school or work. Work through the same steps that Miriam did. What areas are still fuzzy? Review the appropriate sections.

_____

_____

_____

_____

_____

_____

_____

2. Select one or more of the introductory stories from another section of this book. Take it through the same steps that Miriam used. Can you identify all the steps and statistical tools you would use to solve the problem?

_____

_____

_____

_____

_____

_____

_____

FINAL THOUGHTS

You, like Miriam, can know "what's going on out there." You have just studied the same tools she used. You have learned basic modeling and research skills, and you know how to analyze frequencies, discover associations between concepts, and find significant differences among groups.

To make this information even more valuable, consider some additional steps.

1. For your job, for a school research project, or for another important activity in your life, create a database of key measurements. Identify the dozen most important measures of what you do or what you would like to achieve, and update them regularly. Create summarizing numbers that track your activities.

2. Create a spreadsheet template. Use the tools in this book as a start. For each statistical tool you add, make notes about the restrictions that apply: the type of data, the number of cases required, etc. Highlight places where you can enter raw data that the rest of the spreadsheet will grab and process. Add statistical tools from other sources to your template. Make it your own personal expert system, that grows with you, that you can use to analyze the data you confront in your professional life.

3. If you are doing a great deal of analysis, consider investing in dedicated statistical software. It is an excellent way to increase your knowledge of statistical tools, and makes complicated analyses extremely easy.

4. Follow up with higher-level text and courses. You have already studied the fundamental ideas: summarizing numbers and probability distributions. With this knowledge, additional statistical techniques will be much easier to learn.

5. Find ways to apply what you have just learned. The stories and exercises in each section may have suggested ways you could use statistics at school or at work. Select a low-risk environment where you will have time and opportunity to polish your skills, gain confidence, and make mistakes without penalty.

With a little practice, and the success and failures that accompany it, you will understand complicated problems more easily. You will tease useful information from muddled situations. Most important of all, you will act when others are only confused. Nothing is more rewarding than finding important, useful insights that neither you nor anyone else has ever anticipated.

And they are all around you.

SELECTED ANSWERS

SECTION 3

Exercise #6, Question 1

a. ratio   b. interval   c. nominal   d. ratio   e. interval or ratio (assumed)
f. ordinal   g. ordinal   h. interval   i. ratio   j. nominal

SECTION 4

Exercise #9

A phone-in poll on controversial issues is biased toward those who feel intensely enough about an issue to respond. Everyone else has no chance to be selected at all. Therefore, the sample is not random and any use must be qualified.

SECTION 5

Exercise #12, Question 2

120	6	20	12.5	10	60	10
42	7	6	7	7	10	2
65	5	13	11	11	28	2
n/a	3	n/a	n/a	red	n/a	n/a
115	5	23	4	2	102	2

Exercise #13, Question 2

8	2.83	8
50	7.07	20
500	22.36	50
1250	35.36	100
3.2	1.79	4

Exercise #14, Question 1

Summation symbol, standard deviation of a population, mean of a population, mean of a sample, standard deviation of a sample, a value from the data set, the variance of a sample, the variance of a population.

Section 6

Exercise #16, Question 1

15.8%, 2. 2%, 68.2%, 52.2%, 97.8% (Your answers might vary by 0.1–0.2 percentage points, due to rounding in different parts of the chart.)

Section 7

Exercise #18, Questions 1 and 2

1. +2, +3, –0.5, –1.5, –3.   2. –1.5, –3, and –2.4 should be circled.

Exercise #19, Questions 1 and 2

1. 17, 20, 90, 83, 25.   2. 90, 85, 76, and 99 should be circled.

Exercise #20

one, one, two, two

Exercise #21, Questions 1–5

1. 15.8%   2. 2.2%   3. 84   4. 2.2% of 1000 or 22   5. 15.8% and 15.8%

Section 8

Exercise #24

Expected Number of Cases:

	A	B
East	16.6	15.4
West	12.4	11.6

Chi-square:

$(10 - 16.6)^2 / 16.6 = 2.62$ $(22 - 15.4)^2 / 15.4 = 2.83$

$(19 - 12.4)^2 / 12.4 = 3.51$ $(5 - 11.6)^2 / 11.6 = 3.76$

Chi-square = 2.62 + 2.83 + 3.51 + 3.76 = 12.7

Degrees of freedom $= (2 - 1) * (2 - 1) = 1$

Since 12.7 is more than 3.8, these results indicate a statistically significant relationship beyond the 0.05 level.

Section 9

Exercise #25

21, positive; 15, positive; –21, negative; 4, negative; 16, positive

Exercise #26

$109,648; $15,344; $25,448

Exercise #27

25%, negative; 6%, negative; 56%, positive; 9%, negative; 36%, positive

Section 10

Exercise #30, Question 1

Question	Std. Error	Confidence Interval, 95%	Confidence Interval, 99%
1.	0.51	84 to 86	83.7 to 86.3
	1.83	96.4 to 103.6	95.3 to 104.7
	6.2	487.8 to 512.2	484.0 to 516.0
	0.32	19.4 to 20.6	19.2 to 20.8
	0.82	48.4 to 51.6	47.9 to 52.1
3.	0.06%	n/a	4.8% to 5.2%
4.	0.14	6.2 to 6.8	n/a
5.	0.22	$24.57 to $25.43	n/a

Section 11

Exercise #32

Values of z: 0, 3.21, 2.67. All values in excess of 1.65 are significant beyond the 0.5 level. This means that less than 5 times out of 100 would such results occur, if indeed the actual difference were not as large (or larger) than that suggested.

OVER 150 BOOKS AND 35 VIDEOS AVAILABLE IN THE 50-MINUTE SERIES

We hope you enjoyed this book. If so, we have good news for you. This title is part of the best-selling *50-MINUTE*™ *Series* of books. All *Series* books are similar in size and identical in price. Many are supported with training videos.

To order *50-MINUTE* Books and Videos or request a free catalog, contact your local distributor or Crisp Publications, Inc., 1200 Hamilton Court, Menlo Park, CA 94025. Our toll-free number is (800) 442-7477.

50-Minute Series Books and Videos Subject Areas . . .

Management
Training
Human Resources
Customer Service and Sales Training
Communications
Small Business and Financial Planning
Creativity
Personal Development
Wellness
Adult Literacy and Learning
Career, Retirement and Life Planning

Other titles available from Crisp Publications in these categories

Crisp Computer Series
The Crisp Small Business & Entrepreneurship Series
Quick Read Series
Management
Personal Development
Retirement Planning